LAND AND LIMITS

This critical new analysis explores the impact of an influential idea – sustainable development – on the institutions and practices governing the use of land. A central theme is the paradox that despite increasing attention to sustainability, conflict over land use is as ubiquitous, and as intense, as ever.

The authors challenge many prevailing assumptions about 'planning for sustainability'. After charting the remarkable growth in expectations of planning, they show how attempts to interpret sustainability must lead inexorably to moral and political choices of a fundamental kind. Planning, they argue, is not so much a means of implementing some pre-formed, consensual concept of sustainable development as a crucial arena in which different conceptions of sustainability are defined and contested.

Important themes developed in the first half of the book are carried through into chapters that assess the role of planning in three vital areas: transport, biodiversity and minerals extraction. Challenging conclusions are drawn together in the final chapter, and the potential for planning to provide a space for dialogue about environment and development is explored.

Offering a rigorous discussion of concepts, policy instruments and contemporary planning dilemmas, *Land and Limits* will be essential reading for academics, professionals and advanced students in the fields of environmental policy and land-use planning.

Susan Owens completed her first degree and PhD at the University of East Anglia. She is now Reader in Environment and Policy in the Department of Geography, University of Cambridge and a Fellow of Newnham College.

Since completing his PhD at the University of Cambridge, **Richard Cowell** has worked as a Research Fellow and Lecturer at the Department of City and Regional Planning, Cardiff University.

ROUTLEDGE RESEARCH GLOBAL ENVIRONMENTAL CHANGE SERIES

THE CONDITION OF SUSTAINABILITY
Ian Drummond and Terry Marsden

WORKING THE SAHEL
Environment and Society in Northern Nigeria
Michael Mortimore and Bill Adams

GLOBAL TRADE AND GLOBAL SOCIAL ISSUES
Annie Taylor and Caroline Thomas (eds)

ENVIRONMENTAL POLICIES AND NGO INFLUENCE
Land Degradation and Sustainable Resource Management in Sub-Saharan Africa
Alan Thomas, Susan Carr and David Humphreys (eds)

THE SOCIOLOGY OF ENERGY, BUILDINGS AND THE ENVIRONMENT
Constructing Knowledge, Designing Practice
Simon Guy and Elizabeth Shove

LIVING WITH ENVIRONMENTAL CHANGE
Social Vulnerability, Adaptation and Resilience in Vietnam
W. Neil Adger, P. Mick Kelly and Nguyen Huu Ninh (eds)

LAND AND LIMITS
Interpreting Sustainability in the Planning Process
Susan Owens and Richard Cowell

THE BUSINESS OF GREENING
Stephen Fineman (ed.)

INDUSTRY AND ENVIRONMENT IN LATIN AMERICA
Rhys Jenkins (ed.)

LAND AND LIMITS

Interpreting sustainability in
the planning process

Susan Owens and Richard Cowell

London and New York

First published 2002
by Routledge
11 New Fetter Lane, London EC4P 4EE

Simultaneously published in the USA and Canada
by Routledge
29 West 35th Street, New York, NY 10001

Routledge is an imprint of the Taylor & Francis Group

Typeset in Garamond by Taylor & Francis Books Ltd
Printed and bound in Great Britain by The University Press, Cambridge,
United Kingdom

British Library Cataloguing in Publication Data
A catalogue record for this book is available from the British Library

Library of Congress Cataloging in Publication Data
Owens, Susan E.
Land and limits: interpreting sustainability in the planning process/Susan
Owens and Richard Cowell.
p. cm. – (Routledge research global environmental change; 7)
Includes bibliographical references and index.
1. Land use–Great Britain–Planning. 2. Land use–Environmental aspects–Great
Britain. 3. Land use–Government policy–Great Britain. 4. Environmental
policy–Great Britain. 5. Transportation and state–Great Britain. 6. Natural
resources–Government policy–Great Britain. 7. Sustainable
development–Great Britain. I. Cowell, Richard. II. Title. III. Routledge
research global environmental change series ; 7.
HD596 .O938   2002
333.73'0941–dc21   2001040372

ISBN 0–415–16276–9

CONTENTS

ACKNOWLEDGEMENTS

The idea for this book first materialised in 1994, when Susan Owens held an ESRC Global Environmental Change programme fellowship and Richard Cowell an ESRC postgraduate award. Our first debt of gratitude, therefore, is owed to the ESRC for funding the work that originally stimulated our ideas, and especially to Michael Redclift (then director of the GEC programme) and Alister Scott, who encouraged us to produce the book proposal.

At that stage, the project seemed more straightforward, and more tractable, than it eventually turned out to be. The more we thought about the issues of 'planning for sustainability' the more difficult and complex they seemed to become, but we learned a great deal during the process of writing and the result is a more critical book than the one we originally envisaged. Many colleagues helped in this process, and their perceptive comments on successive drafts contributed much to the final manuscript: we are especially grateful to Bill Adams, Michael Banner, Olivia Bina, Marguerite Camilleri, Alison Johnson, Andrew Johnson, David Lewis, Peter Owens, Tim Rayner, Joe Smith and Bhaskar Vira. Our thanks must also go to the supportive editorial staff at Routledge – most recently Simon Whitmore, Joe Whiting and Annabel Watson – who have waited patiently for the manuscript, and to Jane Robinson and Colin MacLennan in the Geography Department library at Cambridge, for their courteous assistance with numerous bibliographical queries.

1

OLD CONFLICTS AND NEW IDEAS

We have failed to see how our economy, our environment and
our society are all one.

UK Prime Minister Tony Blair 1999: 3

not all truths can fit into one social world.

Rawls 1993: 197, citing Berlin 1969

Introduction

Conflict over the use of land often seems ubiquitous. As we write, there are
many examples to choose from in Britain, but a few will suffice to illustrate
the range of issues at stake. In Berkshire, a major company wishes to build
its new headquarters in a controversial green field location. Opponents raise
the spectre of congestion and pollution and claim that the development
would make a mockery of government planning guidelines seeking to
reduce car dependency (Hetherington 1999). The local authority, alarmed at
the prospect of a major employer going elsewhere, and influenced by the
company's 'green transport' plan, grants planning permission after 'acrimonious' debate (*ibid.*: 14; see also Groom 1999a).[1] In Scotland, development
interests propose a funicular railway to carry increasing numbers of tourists
into the Cairngorms, the UK's most significant mountain plateau, justifying
the project – and around £12 million in public assistance – as a boon to the
local economy. Environmentalists fear the impact of more visitors in a
remote and ecologically vulnerable landscape. Planning permission is
granted, but the decision, along with the case for public finance, is then
subjected to challenges from environmental groups and a National Audit
Office review.[2] At the other end of the country, plans for a by-pass that
would bisect the renowned water meadows of the cathedral city of Salisbury
run into deep controversy, undiminished by a public inquiry and a recommendation that the project should go ahead. The government's advisory
bodies on landscape and nature conservation, consulted about ways of mitigating the impacts of the new road, agree that no measures could effectively

1

do so: the scheme is returned to the drawing board.[3] Expanding our focus to the regional scale, we see Britain's 'most important planning inquiry for a generation ... bring[ing] to a head the battles between environmental protesters and developers' in the south-east of England (Groom 1999b: 12). Draft regional planning guidance, produced by a grouping of local planning authorities, seeks to improve environmental quality by such measures as encouraging people to drive less, increasing woodland coverage and deflecting growth pressures from the fastest growing parts of the region (SERPLAN 1999). Industry complains that the guidance embodies 'a presumption against economic development', the Council for the Protection of Rural England (CPRE) that too many new houses are proposed, and the House Builders' Federation that restrictive policies will increase house price inflation, exacerbate labour shortages and threaten international competitiveness (Groom 1999b: 12).[4]

Many characteristics of these conflicts would have been familiar to Roy Gregory, whose classic book, *The Price of Amenity* (Gregory 1971), explored some of the great planning controversies of the 1960s: the claims of growth and jobs set against concerns for less tangible environmental qualities; conflicts between different sets of public and private goals; and issues of distributive justice. Much has also changed. Whereas Gregory charted conflicts over projects promoted by nationalised industries 'in the public interest', the cases above centre upon more complex mixes of private and public development in the context of a more internationalised economy. The nature of environmental concerns has also shifted dramatically, with 'amenity' stretched to include invisible but potentially global threats, such as climate change and loss of biodiversity. What is most striking, however, is the persistent emergence of conflict over the use and development of land, apparently at odds with the more positive, less adversarial future for environmental politics heralded by widespread commitments to sustainable development. It is this paradox that forms one of the central themes of our book.

Of the rise to prominence of 'sustainable development' it is hardly necessary to say very much here: this history has been thoroughly documented elsewhere (see, for example, Adams 1990; Elliott 1999; Holdgate 1996; Redclift 1987). Famously crystallised by Brundtland, sustainable development drew together ecological imperatives, the continuing need for development and concerns for global and inter-generational equity (WCED 1987).[5] Promising to 'transcend national differences and political interests' (Yearley 1996: 100), it resonated with the themes of ecological modernisation already gaining currency in academic and policy communities, particularly in its insistence on the interdependence of environment and economy.[6] Most significantly, sustainable development seemed to move the debate beyond 'limits to growth' and therefore to mark a departure from the old forms of conflict between development and conservation. Its multifaceted nature allowed a wide range of governmental, commercial and voluntary organisations to embrace it and support its use in guiding future policy.

However, apparent universality of appeal did not lead to singularity of definition or noticeable coherence of approach. Competing interpretations of 'need' and 'development', and difficulties in delineating the ecological and social conditions for sustainability, quickly fractured the consensus. Nowhere did the schisms become more visible than in attempts to define principles and apply them to specific activities or environments – precisely where sustainability had to be interpreted if its appeal was to amount to more than 'rhetorical genuflection' (Blowers 2000: 371).[7] By the mid-1990s, many felt that sustainable development had become 'a much abused term' (Quinn 1996: 20) that proponents accused each other of hijacking. The finger was pointed at business interests for gilding their 'green image' (Blowers and Glasbergen 1995: 164); at the rich countries for seeking to sustain their dominance (Yearley 1996); at environmentalists for adopting 'morally repugnant' interpretations (Beckerman 1994: 191, 1995); and at government technocrats for emphasising their own claims to expertise over broader political agendas (see discussion in Rydin 1995). Borrowing a distinction made by John Rawls in his theory of justice, we might argue that the *concept* of sustainable development – the broad meaning of the term – became widely accepted, but the *conception*, which includes the principles required to apply a concept, remained (and remains) in profound dispute (Rawls 1972, 1993; see also Adams 1990; Jacobs 1999; Redclift 1987).[8]

Yet even if sustainable development fails to provide a straightforward 'figure of resolution' (Myerson and Rydin 1996: 26), it would be difficult to deny the discursive power of an idea that, in less than two decades, permeated policy agendas so as to become almost ubiquitous. The fact that Brundtland could promulgate the concept with such effect suggests that it must have been, in some sense, an idea whose time had come. This raises a number of interesting and important questions. How does an idea come to have influence in this way? How is it transmitted and developed? And to what extent can 'the knowledge-based power of persuasion', in which 'ideas and discourses themselves may be powerful entities' (Litfin 1994: 18, 15), challenge more traditional forms of domination and control? Such questions, in a more general sense, are of growing interest to scholars and policy makers. There is increasing evidence that traditional forms of policy analysis that emphasise agency, structures or institutions fail to capture the important role of knowledge, ideas and argument in the political process (see, for example, Fischer and Forester 1993; Hajer 1995; Kingdon 1995; Litfin 1994; Majone 1989; Radaelli 1995; Rein and Schön 1991; Sabatier 1987, 1998; Sabatier and Jenkins-Smith 1993; Weale 1992; Weiss 1991); at the same time, power (in the coercive sense) is seen by many as integral to the legitimation of particular arguments and beliefs (Flyvbjerg 1998). Exploring the influence of a 'new' concept like sustainable development, which in fact has long antecedents (Owens 1994), may throw further light on the cognitive aspects of policy making. On the one hand, we might expect interpretations of sustainability – the emergence of particular conceptions – to be constrained by existing institutions, so that the new idea could be absorbed seamlessly with little threat to the *status quo*; alternatively, we might postulate that the very process of interpretation could stimulate institutional learning, leading ultimately to meaningful and lasting policy change.[9]

We have chosen to explore these issues in the context of policies concerned with the use and development of land. Not only do such policies demand the interpretation of sustainable development in terms applicable to hard decisions but, as we show in Chapter 2, the various processes of land-use planning and regulation have repeatedly been identified as key instruments for delivering a more sustainable society. In many respects, and in many countries, it is land-use planning that has acquired the political burden of reconciling development pressures with environmental concerns. We now turn, therefore, to a consideration of the significance of land in relation to sustainable development before looking briefly at the role of planning and identifying some of the more specific issues to be addressed in this book.

Land, environment and sustainability

As a source of sustenance, resources and wealth, land 'is literally the base upon which all human societies are built' (Caldwell and Shrader-Frechette 1993: 3). Most of the profound changes in the physical and social conditions of human existence have had important land-use dimensions, including, over a long time span, the shift to sedentary agriculture, the Industrial Revolution, large-scale urbanisation (strongly associated with the rise of formal, public, land-use planning) and the globalising influence of information technology.[10] But land is much more than a material base: it is imbued with diverse, sometimes contradictory, social and cultural meanings – as property, when 'mixed with human labour' in the Lockean sense (Locke 1988), as a wider 'biotic community', deserving of a 'land ethic' (after Leopold 1949), and 'as a vast mnemonic system for the retention of group history and ideals' (Lynch 1960: 126). Not surprisingly, therefore, land has been the subject of persistent political struggle (Caldwell and Shrader-Frechette 1993; Grove-White 1991; Whatmore and Boucher 1993), and we might expect the governance of land-use change to present a challenging set of issues for the theory and practice of sustainable development. Certainly, land presents problems for the more sanguine, eco-modernist interpretations of sustainability, which see salvation in improving eco-efficiency, or 'doing more with less'. While it is demonstrably possible to reduce the energy or materials intensity of the economy (and thus the environmental impacts of growth), relationships between economic activity, land and environmental change make analogous assertions about land-use intensity much more problematic (Owens and Cowell 1994).[11] Theorists of ecological modernisation have tended to focus on pollution and resource consumption while saying little about 'intuited nature' (Mol 1996: 315; see also Christoff 1996) – the spiritual, aesthetic and intrinsic qualities of the non-human world that are so often central to conflicts over the use and development of land.

Part of the difficulty is that changes in land use are linked to environmental change through a multiplicity of direct, indirect, sometimes cumulative and often uncertain effects. These links operate at different scales and have economic, legal and political dimensions: conflicts arise between multiple rights and jurisdictions, which ecological science alone can rarely if ever resolve. Interactions between neighbouring land uses have focused the minds of scholars for centuries, as 'nuisances' of noise, odour or pollution have spilled beyond properties and sites (Agricola 1950; Coase 1960). Later came the recognition that impacts can cross national boundaries, as, for example, when land-use decisions in one country influence air or water quality in another. In the twenty-first century, 'the issue of global climate change makes all nations neighbours' (Caldwell and Shrader-Frechette 1993: 162), as, one might add, do other transnational concerns in which land use is deeply implicated, such as loss of biodiversity. In many instances, the growth of global agreements and international organisations concerned with environmental issues has modified traditional assertions of national sovereignty over land resources, already weakened by the effects of ecological and economic processes operating beyond state or local control. In the case of the Cairngorm funicular project, for example, environmental objectors were able to point to the international significance of the summit habitats and the risks posed to montane plant and bird communities by global warming. But internationalisation presents strong countervailing forces: communities can be reluctant to subject mobile economic capital to restrictive land-use policies, as illustrated by business responses to SERPLAN's regional guidance, discussed above.

If land use is of such significance for sustainability, it seems appropriate to focus attention on the planning and regulation of land-use change. All modern societies engage, with varying degrees of enthusiasm, in some such activity, and in a number of countries land-use planning has come to be accepted as an instrument for co-ordinating economic, environmental and social policies (Healey and Shaw 1994). It was an obvious candidate, then, for a central role in the delivery of sustainable development, a role widely endorsed by governments, diverse interest groups and land-use planners themselves. According to some observers, the challenge has been taken up with noticeable effect. Meadowcroft (1997: 167), for example, claims that planning for sustainability is now 'a real world activity of officials and ministries', encompassing the restructuring of traditional planning activities and genuinely novel developments. In the chapters that follow, we examine such claims, taking a critical look at interpretations of sustainable development in planning – their conceptual underpinnings, the instruments available and their manifestations in practice. We ask, in relation to the broader themes of knowledge, power and institutional learning identified above, whether sustainability has offered – or seems likely to offer – anything genuinely novel or radical. Has it, for example, proved its integrative potential in reconciling growth and 'amenity', or fulfilled its promise as an inclusive, participatory ideal? Recognising that there might be a two-way process, we consider not only whether the sustainability agenda has changed planning but also how the particular dilemmas encountered in the planning process have informed the sustainability debate. Thus the interface of sustainable development and land-use planning becomes the lens through which we address wider questions about the impact of a 'new' idea on established institutions, policies and practice.

Our main focus is on land-use – sometimes referred to as spatial or physical – planning.[12] In the UK, from which we draw much of our analysis, this means an emphasis on 'development' as defined in town and country planning legislation, which largely excludes the land-use changes associated with agriculture or forestry.[13] But it is abundantly clear that planning in this relatively restricted sense cannot be treated in isolation from other aspects of environmental policy, including, for example, pollution control, the management of water resources and waste, and nature conservation. Not only must these different aspects interact, but the boundaries themselves have also long been matters for dispute (see, for example, CPRE 2000a; Healey and Shaw 1994; Miller 1993), with tensions highlighted afresh by the demands of sustainability. For many commentators, the very nature of environmental systems requires 'an integrated and holistic approach' (Evans 1997:5), linking land-use planning to other aspects of environmental protection and to policies in sectors such as energy and transport, which often seem to conflict with planning objectives. Certainly, as we shall show in chapters on transport, nature conservation and minerals, we must look to a wider context to understand the capacity of planning systems to respond to new ideas and deliver particular goals. Where ecological and economic processes operate at scales beyond the reach of local, territorially defined regulation, important questions arise as to how planning can be a proactive force in the quest for sustainable development. Throughout, we seek to avoid an insular, technical account of 'land-use planning', remaining sensitive to the porosity of the system's boundaries, the limitations of its traditional remit and its relatively modest part on a wider political stage.

The remit of land-use planning

There are more fundamental questions about the proper role for planning in general, and regulation of the use of land in particular. For some, land-use planning is a justifiable infringement of private property rights only in so far as markets fail because of the well-known problems of externalities and public goods. Beyond this, intervention becomes at best a distortion of market forces and a threat to competitiveness, at worst a form of coercion. Essentially, in this view, the market is seen as the institution that will best satisfy consumer preferences and lead to improvements in welfare.[14] The role of planning is to correct for conditions where land prices provide an imperfect guide to land use, and thus to enable markets to operate more effectively (although some would question its capacity to do even this; see, for example, Pennington 1999). For other commentators, however, planning can and should do more than mop up after market failures. Through its commitment to public engagement (however imperfect the current arrangements), it could provide a forum for dialogue in which citizens, collectively, might choose outcomes that differ substantially from those reflecting the aggregation of consumer preferences (Sagoff 1988). Many see (idealised) planning as a dialogical institution, aspiring to the kind envisaged by Habermas (1986, 1987), one that provides 'a space for conversation between competing conceptions of the good' while meeting the requirements of pluralism that such a space be neutral between those conceptions (J. O'Neill 1998: 18). Such 'communicative rationality' has been a clear influence on planning theory, reflected in the work of a number of prominent thinkers (see, for example, Forester 1999; Healey 1993, 1997, 1998a).

But there are other, important, ways in which the role of planning can be conceived. A moment's thought reveals that for many of its proponents (and in practice), planning has always been more than a facilitator of markets or a neutral forum for dialogue. Rather, it is seen as an institution for promoting particular ends, either through some concept of the 'public interest' (identified by McAuslan (1979), alongside private property, as one of the dominant ideologies of planning law), or by providing a different kind of dialogical space in which particular conceptions of the good might be fostered. Such a view clearly presents a challenge for value-neutral models of professional activity, the image towards which many planners have been drawn (Thomas 1994). And none of the different perspectives outlined so far precludes an analysis of planning as a process through which power is exercised, both visibly, in defence of identifiable interests (Sandbach 1980), and more insidiously, 'masked as forms of truth and knowledge', after Foucault (Richardson 1996: 281; see also Flyvbjerg 1992, 1998; Foucault 1982, 1990).[15] Some have suggested that 'the turn to argument' (Richardson 1996: 279) renders the planning process particularly vulnerable in the latter sense, so that planners, rather than being facilitators in a process of rational debate, become (conscious or unconscious) agents of normalising and disciplinary power.

These different perspectives on the legitimate remit, and actual role, of planning are of crucial importance for conceptions of sustainable development. Fundamental issues emerge in the post-Brundtland era as consistently as they did in the 'growth versus environment' contests described in

Gregory's *Price of Amenity*. Central to different claims about sustainability (as we shall show in Chapter 3) are questions about the status of different needs and demands, the extent to which preferences should be accommodated or shaped, and the roles of agents and institutions in making the relevant choices and decisions. A recurrent dispute, involving all of these themes, is whether the role of planning – and latterly concepts of sustainability – should be 'essentially subordinated to the precepts of growth' (Blowers 2000: 374). We shall argue in later chapters that judgements of value – of what is right and what is good – are unavoidable. Indeed, in contrast to those who seek to minimise the 'subjective' element in decision making, we see them as highly desirable. How such judgements should be made then becomes a central question. Most conceptions of sustainable development have involved a dialogical model, seeing markets (even 'corrected' ones) as inadequate and stressing the importance – indeed the centrality – of public engagement in determining what is sustainable. Setting aside for now the vexed issues of subsidiarity and conflicts of interest between different scales (Dobson 1998; Saward 1993), this would seem to point towards a deliberative rather than a purely instrumental role for planning. However, there remains a deep ambiguity between process and outcome. A vision of planning as a neutral forum for arriving at consensus about policies (Blowers and Evans 1997) sits uneasily with the strong 'outcome ethic' (Jacobs 1997a: 227) that permeates much of the literature on sustainability, in which forms of development that are 'more sustainable' are clearly to be preferred and promoted over those that are not. In practice, as we shall show, the outcome cannot readily be separated from the process: planning is not so much a mechanism for implementing sustainable development as an important forum in which different interpretations come to be contested and defined. There is no prior *conception* of sustainability (as opposed to the broader, consensual concept) independent of this process.

Planning in the UK

Where our discussion is concerned with institutions, practice and specific planning conflicts, we draw primarily upon experience in the UK (for detailed accounts of the planning system(s) in the UK, see Cullingworth and Nadin 1994; DoE 1997a; DETR 1999a, 1999b; Hall 1992; Rydin 1998a; for a view from the United States, see Callies 1999). While inevitably involving some sacrifice of breadth, this national focus helps to ground our argument and to give it coherence across chapters. The focus is not exclusive: we recognise the transnational character of epistemic communities in our field of inquiry and take account of European and wider international influences on UK policies. We also comment upon planning issues, institutions and policies in a range of countries where these provide useful illustrations or interesting contrasts. But we lay no claim to an overview or comparative analysis; nor have we attempted any systematic selection of case studies from different national contexts. Approaches to sustainable development in different planning systems will reflect administrative and legal traditions, ecological problems and political cultures. Acknowledging the important contribution of comparative reviews in both land-use planning and environmental policy (see, for example, Boehmer-Christiansen and Skea 1991; Booth 1996; Davies *et al.* 1989; DoE 1996a; Hanf and Jansen 1998; Jänicke and Weidner 1997) but conscious of the pitfalls of such an approach given our subject, we have opted for more limited geographical scope. Our primary concern is with the underlying themes, conflicts and policy dilemmas, which, we suggest, transcend national boundaries and resonate across quite different political cultures and administrative systems.

However, nowhere has the idea of sustainable development emerged or taken hold in a vacuum. Since we draw extensively on experience in the UK, it is worth noting at the outset some important aspects of the British political and economic scene that had particular significance for the way in which planning systems were able to respond to the challenge of sustainability. The most important of these was the neo-liberal economic agenda, with its emphasis on deregulation and privatisation (Blowers 2000). At its most intense in the 1980s, this ideology retained force into the following decade and was not substantially diminished by the election of a Labour government in 1997. While not unique – many other countries have pursued the goal of market liberalisation – these commitments have been particularly marked in the UK. The planning system was itself a direct target for deregulatory zeal in the 1980s, when it was seen as part of the 'burden' that it was the duty of governments to 'lift' (UK Government 1985; Allmendinger and Thomas 1998). Renewed emphasis on the presumption in favour of development (DoE 1980; Miller 1999a) had some significant impacts on the ground (notably in the proliferation of out-of-town retail facilities), but apart from the introduction of enterprise zones and simplified planning zones, in which development would not require planning consent, effects on the *institutions* of planning were surprisingly limited. Indeed, the tide seemed to turn with the introduction of a plan-led system in the Town and Country Planning Act of 1990. Nevertheless, concern about the adverse impacts of planning regimes on competitiveness persisted in the 'modernising planning' agenda of the late 1990s alongside, and indeed influencing, conceptions of sustainable development (DETR 1998a; McKinsey Global Institute 1998).

Also significant were powerful centralising forces operating under successive Conservative administrations and the corresponding squeeze on local government finances. As much as the ideological attack on planning, this withdrawal of resources left local planning authorities in a demoralised position by the end of the 1980s and goes some way towards explaining the enthusiasm with which many planners embraced the sustainability agenda.

As we shall show in Chapter 2, the recognition that sustainable development needed land-use planning, and the empowerment that this promised, can be positioned alongside a more general (if qualified) resurgence of interest in planning and positive regulation (Majone 1996). After 1997, the context for planning shifted again, with devolution of powers to the Scottish Parliament, National Assembly for Wales and Northern Ireland Assembly, and increasing emphasis on the regional level of decision making in England. One effect is that planning for sustainability in the UK is likely to become an increasingly differentiated affair.

An outline

The next chapter provides an overview of commitments to sustainable development from international to local levels, and of the prominence accorded to planning as a means of converting this rhetoric into action: its aim is to set the broad policy context. Chapter 3 then turns to the conceptual basis for sustainable development, beginning with models grounded in concepts of environmental capital and exploring variants of these and alternative interpretations. We show that attempts to move beyond Brundtland to introduce rigour and specific decision rules expose fundamental moral and political challenges, and we suggest that it is these basic tensions, rather than the absence of an unambiguous definition, that makes planning for sustainable development less tractable than some had expected. Chapter 4 returns the focus to planning practice, with a critical examination of selected policy instruments. In particular, we look at the propensity of different approaches to promote technical rationality or a deliberative agenda and, in turn, to favour particular conceptions of sustainability.

Our arguments are then developed in three further chapters in which we are concerned with interpretations of sustainable development in specific policy sub-systems, all implemented at least in part through the land-use planning process. Each is chosen for its capacity to highlight fundamental questions that planning for sustainability raises but cannot, of itself, resolve. Chapter 5 addresses the potential to promote a sustainable land-use and transport system through locational policies, a much vaunted role for planning that, we argue, exposes tensions between accommodation and management of demands, and between planning as a means of promoting efficient outcomes (in the neo-classical economic sense) and planning as a forum in which preferences might be shaped and modified. Chapter 6 turns the attention to nature conservation, and to the conceptual and ethical difficulties that must be negotiated in generating defensible arguments and procedures for habitat protection: it shows very clearly that judgements of value are indispensable to the project of sustainable development. In Chapter 7, our focus is on issues of governance and spatial equity, addressed in the context of minerals planning and the problems of interpreting sustainability in ways that are responsive to diverse local economic and environmental conditions.

We could have chosen to explore these issues within the context of any number of different policy areas with an important land-use dimension — housing, energy, water and waste management are prominent contemporary issues that spring to mind. However, while the specific challenge of sustainability in individual sectors is interesting in itself, the dilemmas revealed in the process of land-use planning have broader applicability precisely because of the fundamental nature of the tensions thereby exposed. We draw this material together in Chapter 8, where we revisit the core themes of the book and the questions identified in this introduction.

2

RHETORIC, POLICY
AND PRACTICE

Sustainable development as a planning issue

The planning system, and development plans in particular,
can make a major contribution to the achievement of the
Government's objectives for sustainable development.

DETR 1999b: 20

Usable doctrines do not spring, fully armed, from a theorist's
brow. They have to be hammered out in the give and take of a
debate, provoked and shaped by the lived experience of particular societies at particular times.

Marquand 1988: 12

Introduction

In his celebrated account of the 'issue-attention' cycle, Anthony Downs
(1972) suggested that after the initial discovery of an environmental
'problem' and a stage of euphoric enthusiasm, proponents would come to
realise the difficulties and costs of making significant progress. In many
cases, this realisation would be followed by a gradual decline of public and
political interest, and new concerns would then take over. A small number
of issues would survive, however, to become matters for political attention or
decision. Why some issues are consigned to oblivion while others stay firmly
on the agenda has been a question of considerable interest to political scien-
tists and policy makers (for an early but still fascinating 'insider' account,
see Solesbury 1976; see also Kingdon 1995; Hajer 1995). Such an attention
cycle might also be postulated for concepts and ideas, and similar questions
asked about their durability.

Our intention in this chapter is to tell the story of the diffusion of a new
idea during the decade or so after its popularisation by Brundtland. We
show how sustainable development took hold as a powerful normative and
intellectual framework for land-use planning, from a period of euphoric
enthusiasm at the start of the 1990s to more sober assessment as difficulties
and complexities became apparent; and we note that, in many respects, this
transition has been a microcosm of experience across a wider range of policy

areas. Our intention here is to provide background and to formulate questions, and thus to set the scene for a more analytical approach to selected issues in the remainder of the book.

We begin with a critical review of the endorsements heaped upon planning as an instrument of sustainable development and suggest that the range and potentially contradictory nature of the issues involved was always likely to overburden planning systems, in the UK and elsewhere, with unrealistic expectations. We then show that while sustainability found its way quite rapidly into formal statements of policy at all levels, significant discrepancies emerged between rhetoric and aspiration on the one hand and practice and implementation on the other. Such a shortfall, we contend, cannot simply be interpreted as an 'implementation gap' – a natural time lag in the application of new but broadly consensual principles. Rather, it reflects a struggle to *interpret* sustainability in which the process of definition – and not just 'implementation' – becomes integral to the politics of land-use change.

Sustainable development: the need for planning

The rapidly diffusing attention to sustainable development coincided with, and promoted, a resurgence of interest in national and land-use planning. This can be traced in part to the international sphere, as first Brundtland in 1987, and then the 1992 United Nations Conference on Environment and Development (UNCED), took the view that addressing the interrelated problems of environment and development demanded the production and implementation of strategies or plans. A commitment to national sustainability plans was a key component of the UNCED agreements, while the main statement of principles for action, Agenda 21, regarded 'environmentally sound physical planning' as essential to sustainable development in urban areas and desirable in the integrated management of land resources elsewhere (UNCED 1992a: para 7.28; see also Selman 1996). At the European level, the Fifth Environmental Action Plan (CEC 1992a), *Towards Sustainability*, afforded planning a significant role in shifting from reactive responses towards anticipatory and strategic resolutions to environmental problems.[1] Later, the European Union began to promote a more extensive role for spatial planning at all scales in pursuit of its agendas for the single market, social cohesion and environmental sustainability (Alden and Boland 1996). In 1999, it adopted the European Spatial Development Perspective as a legally non-binding framework for better integration of all sectoral policies with territorial effects (CEC 1999a). However, we should also note that non-governmental organisations (NGOs), such as the International Union for the Conservation of Nature (IUCN), have helped to highlight the importance of comprehensive and environmentally informed planning systems (IUCN *et al.* 1991; Bishop 1996).

These ideas about planning and sustainability found a receptive audience, especially among local governments. In the UK, as elsewhere, they coincided with (and reinforced) a wave of environmental concern, affecting 'urban areas as much as rural, and local councils of all political persuasions' (Hall *et al.* 1993: 24; see also Bain *et al.* 1990; Ward 1993). By the start of the 1990s, about three-quarters of councils in England, Wales and Scotland already had a 'green plan' or charter of some kind in effect or in preparation (Wilson and Raemakers 1992), some explicitly recognising the need to extend consideration to 'issues such as global warming, ozone layer depletion, and destruction of the rain forest, as well as local matters' (London Borough of Hounslow 1991, cited in Ward 1993: 460). Thus connections between planning and emerging concepts of sustainability were often forged in the light of earlier initiatives that for the most part lay outside the statutory domain. What was new in the 1990s was the degree to which central government began to give formal support to sustainable development, and to land-use planning as a means of achieving it, in some cases producing the relevant policies and legislation in response to international commitments. Indeed, in many countries, legitimation by central government was crucial in allowing innovation at the local level to become formalised in plans and strategies. In the Netherlands, spatial planning came to have a well-defined role in the National Environmental Policy Plan, in which environmental targets gave form to interpretations of sustainability (Netherlands Ministry of Housing, Spatial Planning and Environment 1989, 1990, 1994, 1998).[2] In Norway, the need to promote sustainable development was a major factor in the *instigation* of national planning policy guidance, and planning in a more general sense has been expected to co-ordinate activities concerned with the use and protection of resources alongside balanced and equitable regional development (Kleven 1996; Sverdrup 1998). There has been similar recognition of a role for planning in promoting sustainable development in Canada (Perks and Tyler 1991; Richardson 1989; Tomalty and Hendler 1991), Germany (Marshall 1992), New Zealand (Memon and Gleeson 1995; Southgate 1998; Upton 1995), Sweden (Elander *et al.* 1997) and (to some extent) the United States, where the relationship has been manifest in the 'smarter growth' movement.[3]

The UK government's first official endorsement of sustainable development, in the Environment White Paper of 1990, urged planning authorities to 'integrate environmental concerns into all planning policies' (UK Government 1990: 65), and a statutory basis was provided by the Town and Country Planning Act of the same year, which required all development plans to include policies for conservation of 'the natural beauty and amenity of the land', improvement of the physical environment and the management of traffic.[4] In the first national sustainability strategy for the UK, planning was seen as 'a key instrument' for delivering sustainable land-use change (UK Government 1994a: 221), an emphasis reflected (if not always coherently) in subsequent revisions to planning policy guidance notes (PPGs), which interpret, for local authorities, government policy concerning the role of the planning system in a range of different contexts. Significantly, a revised version of the guidance setting out general policy and principles established sustainable development as the first of three themes 'which underpin the Government's approach to the planning system' (DoE 1997a: para 3). Beyond these core policy documents, endorsements for planning featured regularly in ministerial statements, speeches[5] and evidence to parliamentary select committees (notably the House of Lords Select Committee on Sustainable Development 1995), and survived the 1997 change of government largely intact, although with some shift of emphasis. The British government has generally fought shy of making sustainable development a legal requirement, but exceptions, outside the town and country planning system itself, include the Natural Heritage (Scotland) Act 1991 and the Government of Wales Act 1998. Section 121(1) of the latter charges the National Assembly for Wales with a duty to 'make a scheme setting out how it proposes, in the exercise of its functions, to promote sustainable development'.

The interest of the planning profession itself was reflected in a profusion of conferences and reports trying to make sense of sustainable development (see, for example, Blowers 1993a; County Planning Officers' Society 1993; Welbank 1993). Initial confidence that planning should have a prime role in promoting sustainability seemed hardly diminished by confusion surrounding the concept and its implications for land-use change; in terms of Downs' issue-attention cycle, the profession was caught up in the stage of euphoric enthusiasm in the early part of the 1990s, and the impetus was carried forward into statutory planning processes as a growing number of local authorities moved beyond rhetoric to integrate the new concept into their development plans. We consider the efficacy of these efforts in more detail later.

Planning also attracted renewed attention from other organisations. Statutory conservation bodies were quick to draw connections between sustainable development, land-use planning and their own objectives. They endorsed the view that 'the strategic planning process can play a key part in moving society towards the idea of sustainable development' (Countryside Commission et al. 1996: 4) and continued to develop and stress these links (Collis et al. 1992; Countryside Commission 1996, 1998; Countryside Commission et al. 1993, 1996; David Tyldesley and Associates 1994; National Rivers Authority 1994). Pressure groups that had long regarded the planning system as an important lever on environmental change were also quick to seize new opportunities, sometimes seeming to inundate planners and policy makers with encouragement, information and advice. The Council for the Protection of Rural England (CPRE), for example, forcefully maintained that the planning system had a central role 'in controlling the use of natural resources as part of meeting our local, national and international responsibilities towards the achievement of sustainable development' (CPRE 1991: para 3; see also McLaren and Bosworth 1994; National Trust 1999; RSPB 1993). Less predictable, perhaps, was the acknowledgement of an important role for planning from organisations as diverse as the Rural Development Commission (1995) the Confederation of British Industry (CBI and RICS 1992)[6] and the Automobile Association (AA undated).

We might pause to consider why such diverse institutions seized upon land-use planning as a means of promoting sustainable development, using remarkably similar language (Myerson and Rydin 1996). That some strange bedfellows were able to endorse sustainable development at all demonstrates the versatility of the discourse and the ability of different groups to construct it in their own image. Planning was important in this context, as an arena in which favoured interpretations of sustainability could be directed and shaped, ensuring that particular environmental qualities, social practices or economic strategies could survive. In the case of government, we can sometimes detect political expediency and legitimation at work, with suitably vague planning rhetoric seeming responsive to demands for new approaches while deflecting awkward questions about the growth orientation of private investment and public policy. Land-use policies – at least before they bite – can sometimes provide a reassuring institutional alternative to measures with more immediate effect, such as green taxes; and when they *do* bite, local authorities, rather than central government, face the political repercussions. In such circumstances, it is convenient that advocates of sustainable development call for greater local autonomy. Planning undoubtedly has been, and continues to be, used in such ways, as we show in relation to environmental assessment, transport and other issues in later chapters. But it would be too simple to interpret developments purely in terms of expediency and legitimation. Faith in planning as an instrument of sustainability may also follow naturally from its function in 'reconciling' development and conservation, even if territorially rooted systems are less obviously equipped to respond to new and pervasive environmental problems. At some fundamental level, there is an inescapable role for planning if decisions about land use are important in determining what is 'sustained'.

Whatever the rationale, the broad consensus that sustainability needed planning at various levels resonated with – indeed, helped to provide – a renewed impetus for the organised, public direction of social and economic development (Blowers and Evans 1997), co-existing somewhat uncomfortably with the prevailing ideological emphasis on markets and competitiveness. By the mid-1990s 'planning and sustainability' was firmly on the policy agenda. By the end of that decade, the British government could strenuously assert that sustainable development lay at the heart of national policy for the planning system. Probing beyond such formal support, however, reveals uncertainty and confusion about interpretations of sustainability and about the remit of the land-use planning system in its pursuit of this goal. Almost inevitably, the expectations of planning, which we now go on to assess, outstripped the capacity of the system to incorporate new thinking in plans and policies and, more pertinently, to institute policies that were genuinely different from those that had gone before.

Great expectations

One aspect of the growing interest in land-use planning as an instrument of sustainable development has been the expanding range of social, ecological and political objectives that planning systems are deemed capable of promoting. Seen collectively, these aspirations amount to a potentially overwhelming brief for institutions with something of a mixed record in meeting public and political desires (Cullingworth 1997). A brief review of these expectations is useful for two reasons: it provides some insight into the different interests assembled under the banner of sustainable development; and it shows that planning in general, and land-use planning in particular, has become an important arena for articulating and mediating conflict about the meaning of the term.

A prominent dimension of sustainable development has always been the requirement to confront global environmental problems, especially those that have become 'emblematic' (Hajer 1995: 7) at any particular time. So, for example, traditional planning concerns in the UK such as green belt, landscape quality and the built heritage were joined by issues such as climate change, consumption of non-renewable resources and the cumulative impact of development decisions on biodiversity. This constituted something of a new remit for planning (Owens 1994; see also Breheny and Rookwood 1993; DoE 1992a), recognising interrelationships between land use, resource consumption and environmental processes extending across space and time.[7] By the end of the 1990s, draft planning policy guidance included a comprehensive list of environmental and natural resource considerations for development plans, by this stage set firmly alongside economic and social considerations, to which we return below (DETR 1999b). Here we elaborate on a selection of these environmental issues, which, although not exhaustive, indicates the interesting mix of 'old' and 'new' concerns with which the planning system has been expected to engage.

First, and in spite of the traditional demarcation of land-use planning and pollution control,[8] local planning authorities have increasingly been engaged in a wider pollution agenda, in some cases following policy developments at national level. Particularly important for its integrative potential has been the national air-quality strategy (DoE 1997b; DETR 2000a), which requires that local authorities set up air-quality management areas in localities where pollution objectives are unlikely to be met by 2005.[9] No additional planning powers were provided in this context, the expectation

being that planners would use the traditional instruments of locational policies (especially to influence traffic levels) and urban design (to encourage the
dispersal of pollutants).[10] Some local planning authorities have begun to test
the extent to which they can influence land-use change on the grounds of air
quality alone (Land Use Consultants 1999), for example by including appropriate policies in their plans and refusing applications for development that
they consider would add to pollution.[11] At the same time, reduction in
greenhouse gas emissions (not covered by the air-quality strategy) has come
to be seen as an important and legitimate consideration for development
plans, with particular emphasis on reducing pollution from transport.
Drawing inspiration from Dutch experience, planning policy guidance on
transport was first published in the UK in the mid-1990s (DoE and DoT
1994), with a clear emphasis on reducing the need to travel and its associated environmental impacts. We consider the significance of this
development in more detail in Chapter 5.

A role also persistently urged upon the planning system, particularly by
environmental groups, has been that of promoting the sustainable use of
renewable and non-renewable resources – including water, minerals and
energy – through appropriate policies and design. Here the desired input of
planners ranges from the strategic level – where, for example, they might
help to identify the capacity of different areas to accommodate emergent
technologies such as wind energy – to the micro scale, where building design
and layout can have a significant influence on energy consumption.[12] Much
of this thinking remains aspirational, however, not least because guidance on
including resource management considerations in development plans stresses
that they 'must be capable of being addressed through the land use planning
system' (DETR 1999b: para 4.2). A role for planning in managing or even
restricting resource use raises further contentious issues about environmental
capacity (to which we return) and constitutes a challenge to the traditional
remit of the system (Owens and Cowell 1996). Many have advocated restrictive planning policies, for example where meeting growing demand for
water might have significant detrimental impacts (Summerton 1998; UKRT
1997a), and the government accepts that water resource availability should
be taken into account in forward planning (DETR 1999b). Interestingly,
however, in his report *Regional Planning Guidance for the South East of England*
(see Chapter 1), Professor Stephen Crow argued that where social and
economic considerations outweigh issues of water availability, 'then it is
simply up to the water companies to do their [statutory] duty cheerfully'
(South East Regional Planning Guidance Examination in Public Panel 1999:
para 9.21). In other cases, as with minerals, the planning system has traditionally been seen as an instrument for ensuring the *availability* of resources
by making sufficient land allocation.

Closely related to these issues is the theme of making careful use of the
land resource itself by encouraging the remediation and reuse of derelict or

contaminated sites while conserving 'undeveloped' areas. Always a key concern – protection of green belts and agricultural land have been long-standing priorities in the UK – the need to recycle land was given renewed impetus in the late 1990s by the publication of controversial household projections (DoE 1995; see also Bramley and Watkins 1995; CPRE 1997; UK Government 1995a, 1996; UKRT 1997b). Policy guidance on the overall purpose of the planning system (PPG 1) sets out urban regeneration and reuse of previously developed land as 'important supporting objectives for creating a more sustainable pattern of development' (DoE 1997a: para 7), and the Labour government elected in 1997 increased its predecessor's target for accommodating new dwellings on previously developed land from 50 to 60 per cent, a target still considered insufficiently ambitious by some (DETR 1998b; UKRT 1997b). A task force set up to identify causes of, and solutions to, urban decline was clear that if this target is to be met, 'we must make best use of derelict, vacant and under-used land and buildings before we develop on greenfield sites', and it proposed a number of planning and fiscal measures to achieve this objective (Urban Task Force 1999: 5).[13] 'To promote more sustainable patterns of development' (DETR 2000b: para 1), revised planning policy guidance on housing reiterated the 60 per cent target, introduced a sequential test giving preference to brown field over green field sites, and placed the emphasis in housing provision firmly on existing towns and cities.

A fourth example is the growing expectation that planning will link development with conservation, by 'safeguarding the natural world' (Breheny and Rookwood 1993: 9; see also DoE 1994a; DETR 1999b) and by 'protecting what is most valuable in the cultural environment' (Countryside Commission et al. 1993). These charges have been influenced, and to some extent reconstructed, by the new emphasis on sustainable development. Concepts of 'environmental capital', for example, have informed policies on conservation, and the rationale for nature protection has been greatly influenced by globalising concepts of biodiversity: we return to these issues in Chapters 3 and 6. Finally, a number of commentators have noted the potential for land-use planning to facilitate adaptive responses to environmental change, for example by regulating development in flood plains and in coastal areas affected by sea-level rise, or by allocating land for the migration of ecosystems adjusting to climate change (for a discussion, see Owens and Cope 1992).

This ambitious environmental agenda dominated the engagement of the planning system with sustainable development during the early 1990s: if planning was to be an effective instrument of sustainable development, it had to take account of the environment 'in the widest sense' (DoE 1992a: para 6.3). In this process, two different but related sets of forces were in action. On the one hand, the holism of sustainability pushed at the institutional boundaries of the (town and country) planning system in a number of ways, raising questions about the adequacy of its traditional instruments and powers and extending engagement (intellectual if not always practical) into sectors where its statutory remit had historically been limited. At the same time, the concept of sustainable development, although rooted in strongly material concerns for poverty and ecological survival, was itself being elaborated to accommodate familiar planning issues of amenity, townscape and culture, in effect becoming part of a more generalised concern to maintain and enhance the quality of life (see, for example, DoE 1994b). Similar developments took place in other countries. In the Netherlands, for example, the theme of 'disturbance', relating to 'environmental problems which threaten the quality of the immediate living environment of the population', has been an important one in the country's National Environmental Policy Plan (Netherlands Ministry of Housing, Spatial Planning and the Environment 1998: 49).

Social dimensions of sustainability were less prominent in Britain during the early 1990s, despite intra-generational equity being a central component of Brundtland's interpretation (WCED 1987). However, this changed when first the European Union, then the Labour government elected in 1997, placed greater emphasis on reducing social exclusion. A second sustainable development strategy for the UK, significantly entitled *A Better Quality of Life*, was adamant that 'development which ignores the essential needs of the poorest people, whether in this country or abroad, is not sustainable development at all' (DETR 1999c: para 4.2). The strategy now added social considerations to the economy/environment axis that had previously dominated interpretations of sustainability, completing a shift in official rhetoric from a broadly environmental agenda (see, for example, UK Government 1994b, 1995b), through reassertion of the significance of growth, to the relative downplaying of environmental considerations in later interpretations.[14] The more explicit emphasis on social inclusion expanded the sustainability agenda for planning in interesting ways. Because of its provisions for public involvement, planning could now be legitimised as a strategically important, if imperfect, set of institutions for inclusive participation and procedural fairness and, given its potential distributive role in relation to health and environmental quality (Blowers 1993b), as a mechanism for substantive aspects of social justice. The social dimension of sustainability became more explicit in planning policy guidance,[15] which urged, for example, that development should 'benefit communities economically, socially and environmentally' (Scottish Office Development Department 1999: para 65), and that planning authorities 'should consider the extent to which they can address issues of social exclusion through land use planning policies' (DETR 1999b: para 4.11).[16] It also began to appear in the criteria and objectives used in the appraisal of development plans (see, for example, City and County of Cardiff 1997; Leicestershire County Council *et al.* 1998).

A new rationale for planning?

It has not been difficult to show that sustainable development, in a wide variety of interpretations, needs planning. But by the end of the 1980s, planning, in Britain especially, had come to need sustainable development at least as much. Associated in the public mind with unpopular urban redevelopment, and blamed by neo-liberals and some business interests for retarding market forces,[17] the profession was in something of a 'marginalised and precarious position' (Selman 1995: 1). We have already suggested that environmental issues in general, and sustainable development in particular, were sources of new momentum and support for planning, leading some to argue that planning could regain a more powerful and visionary role in constructing sustainable social futures (Evans 1997). On the ground, the role of planning in protecting popular amenities was strengthened in many constituencies by 'an alliance of old-style preservationists, concerned with local issues of environmental defence, and new-style environmentalists, concerned with wider issues of environmental quality and "sustainability"' (Cullingworth 1997: 948–9). Local government sought and gained legitimacy, if not necessarily more power or resources, as the centre came to recognise that some aspects of the green agenda could be pursued more effectively (and perhaps more safely) at a local level (Tewdwr-Jones 1996; Wilson 1998).[18]

At the same time, sustainable development helped to reinvigorate arguments for regional planning in the UK and at a European scale (Blowers 1993b; RSPB 1997). By the end of the decade, it was envisaged that regional planning guidance in England, which provides the strategic framework for lower-tier development plans, would be broadened to embrace a wide range of activities related to the use of land, such as regeneration, transport and environmental protection, as well as traditional locational policies that would be implemented through the statutory land-use planning system. In other words, it would constitute a more comprehensive and integrated 'spatial' strategy, and regional planning bodies would be expected to work closely with other interests, including the regional development authorities (RDAs), which became operational in 1999 (DETR 1999a, 1999b, 1999d).[19] The guidance would also be subject to 'sustainability appraisal' (DETR 2000c).

To summarise our argument so far, while the 1980s had seen something of an assault on the domain and remit of public planning, the growth of concern for sustainable development during the 1990s generated important changes in government rhetoric, some of which were backed by policy change. Expectations rose dramatically that planning could contribute to a broad spectrum of environmental, social, economic and democratic objectives. Voices denying a role for planning have been few and far between (Lai 1999; but see Pennington 1999); even so, the range of concerns, some novel, some placing a new emphasis on matters long found to be difficult, has raised questions about the cognitive, regulatory and political capabilities of existing planning institutions. Many commentators have stressed the need to integrate land-use planning with other environmental planning systems, and to co-ordinate the system with pricing and fiscal policy at a national scale (Blowers 1993b; Cullingworth 1997; Owens 1995). Others have called for less emphasis on technocratic regulation and more on the provision of a forum for democratic participation and social mobilisation (Blowers 1997; Davies 1998). We consider attempts to pursue these objectives in later chapters. However, we now turn to the experience of local planning authorities as they tried to meet some of their own and others' expectations and grappled with the concept of sustainable development during the 1990s.

Rhetoric and reality

Widespread endorsement of sustainable development has always been matched by a degree of circumspection as to how it might be achieved. It became customary for academics exploring the issue to set out a bewildering range of definitions (see, for example, Pearce *et al.* 1989), and those involved in planning quickly perceived that the breadth of concerns aligned under this banner might be difficult to translate into coherent policy initiatives. As the County Planning Officers' Society (1993: 37) observed in an early report on the subject, 'the picture is a particularly confusing one because some aspects of the argument involve ... life threatening considerations at a global scale, whereas others are matters of qualitative judgement at a local level'. Nevertheless, as we have shown, almost everyone agreed that planning would be important in the achievement of sustainability, even if they were not quite sure how it would play its part.

One way to gauge the impact of the new idea is to look at the extent to which it influenced the content of development plans, leaving aside for now more difficult questions about implementation. This approach is complicated by the notoriously slow process of revision and renewal of plans, so that their policies may lag behind the latest thinking on issues like sustainable development. Nevertheless, content analysis, especially when it follows the planning process through successive drafts, can be revealing about patterns of institutional learning and interpretations of sustainability. Here we consider experience in the UK, where enhancement of the status of development plans (in the 1990 Town and Country Planning Act) coincided with enthusiasm for sustainable development in the aftermath of the Brundtland Report. But the British experience is far from unique: similar issues have been faced in many countries in the production of spatial or physical plans providing a long-term framework for land use and physical development.

By the middle of the 1990s, the sustainability agenda was making itself felt at the rhetorical level at least: more than half of the structure plans (prepared by counties) in England and Wales included some concept of sustainable development in a published document (Counsell 1998). Initially, many planning authorities used the idea as a vehicle for traditional environmental concerns (Healey and Shaw 1993; Myerson and Rydin 1994; Punter and Carmona 1997), although some embraced elements of the 'new remit' identified earlier in this chapter (Bruff and Wood 1995). In one of the most comprehensive analyses of the treatment of sustainability in development plans, Counsell (1998, 1999a)[20] found that structure plans in England and Wales performed 'best' on the policy areas of wildlife and countryside, land use, built environment and – then a relatively new concern – waste management. The weakest areas were pollution, natural resource management and socio-economic aspects of sustainability. The evident progress in at least some well-defined policy areas was not matched by a grasp of what Counsell categorised as 'key themes and principles' of sustainable development (such as futurity, equity, holism and global stewardship); even the most innovative authorities tended to deal with these concepts 'primarily in a rhetorical manner' (Counsell 1999b: 46).

These patterns should not be surprising, since it is relatively easy to make generalised commitments, such as minimising the threat to the Earth's climate, or promoting the efficient use of energy resources, especially when they can be presented as aspirations with which it would be perverse to disagree. It is more difficult to translate these into policy measures that genuinely signal a change in direction.[21] In pragmatic terms, planning authorities have limited resources, time and expertise to pursue and implement an ever-expanding array of objectives addressing global pollution and the management of resources. A more fundamental question is how robust the resulting policies can be in the face of competing aspirations and institutionalised presumptions about the legitimate remit and powers of planning. The processing of the first round of plans embracing sustainable development provides some indications. Policies seen as unduly restrictive of development (usually based on some concept of environmental limits) have tended not to survive; nor have those that trespass too far beyond the bounds of land-use planning. Counsell's (1999a–c) findings in this context are revealing. In a celebrated case, West Sussex County Council, which had proposed growth-restrictive policies based on an environmental capacity study, was directed by the Secretary of State to increase its housing allocation. So was Bedfordshire, which had adopted a 'precautionary' policy on land allocation on the grounds that household growth forecasts might never be realised; in this case, the policy did not survive the examination in public.[22] A commitment to the precautionary principle in the consultation draft of the Oxfordshire structure plan was moved to 'supporting information' following objections from developers. References to 'social equity' as an element of sustainable development were either modified (Bedfordshire's was replaced, after the examination in public, by 'social opportunity') or had yet to complete the tests of consultation and inquiry.

Also significant in this context is the geographical variation in treatment of sustainable development: in looking for the most comprehensive attempts to make the concept operational, Counsell (1999a: 132) found that:

> The results showed a geographical bias towards the south and south eastern parts of the country – areas which are experiencing the greatest pressures for development and where local politics favour restraint. The structure plans which performed least well, on the other hand, showed a bias towards Wales and the north of England, where political priorities tend to favour growth.

Such spatial variation should not be surprising, since the planning system is unlikely to be a passive translator of broad policy initiatives. If sustainable development is constituted in terms of different goals, networks of local interest inevitably influence how such goals are reconciled in relation to their particular priorities. Thus in Glasgow, a city with high levels of unemployment and deprivation, the guiding principles of sustainable development for the local plan review, a preliminary to the production of a city-wide local plan by the new Glasgow City Council, included presumptions in favour of both job creation and the disadvantaged in society, and a firm emphasis on 'sustainable regeneration' (Glasgow City Council 1998).

The overall picture is one of an unevenly advancing front along which, even when the rhetoric has been impressive, the depth, breadth and *nature* of commitment have varied a great deal. The institutionalised domain of development planning has confined attempts to deal with the 'new remit' primarily to land-use and locational policies; authorities that have tried to embrace a wider environmental or social agenda have been brought rather sharply back into line. It is not surprising, perhaps, that those that have run furthest with the concept of sustainable development, predominantly adopting an environment-led approach, have encountered major problems in reconciling their aspirations with what seem to be inexorable trends in production, consumption and mobility. This fundamental conflict has manifested itself most clearly when plans and proposals have come under scrutiny during the processes of consultation and examination in public. We can conclude that there has indeed been a marked change in planning discourse, but whether this is associated with sufficient power to challenge dominant paradigms of growth is a question for later chapters. For now, we can observe that despite the firm place of sustainable development on the national agenda, and its diffusion into plans and policies at all levels, there was still, by the end of the 1990s, a sense of implementation deficit. Real changes on the ground – to the nature and form of development, or to the intensity of conflict over land use – were not readily discernible, or at least could not unambiguously be identified as sustainability in practice. In this respect, the treatment of sustainability in development plans reflects a more widespread experience in different sectors and at different levels of governance.[23]

One obvious response to impatience is that to expect visible results so soon would be premature. Perhaps what should surprise us is the speed of diffusion and permeation of the idea of sustainable development in little more than a decade. Applying policies in the real world is inevitably less tidy than formulation of abstract principles, and implementation deficits are endemic to public policy as governments seek to address changing social and political priorities in the absence of consensus (Kleven 1996; Weale 1992). In the case of land-use planning, the slow pace of plan preparation results in an almost permanent state of dislocation between new commitments, policy action and land-use change (Quinn 1996; UKRT 1997c). Initially, such structural constraints were compounded by a dearth of practical advice, leading to complaints that planning was being asked to fight the battle of sustainable development 'with no weapons and with very little strategic direction' (Welbank 1993: 13), although gradually this vacuum was filled by revised planning policy guidance together with a profusion of good practice guides (see, for example, Barton *et al.* 1995; DETR 1998c; RTPI 1996, 1999). In Chapter 4, we return to the suitability of various 'weapons' and the ability of the planning system to deploy them to greatest effect. For now, we note two important features of land-use planning that have affected its ability to deliver. First, in common with a number of other regulatory systems, it has limited capacity in isolation to stimulate social and economic change. Second, as we showed earlier, planning authorities were dealing with many other pressing issues while trying to come to terms with sustainability: pressures to speed up the planning process; reductions in public spending; the privatisation of many services; and, from the late 1990s, devolution and the growing significance of the regional agenda (Blowers 2000). Some of these forces were clearly in tension with the thoughtful and inclusive incorporation of sustainable development, some offered new opportunities, but all demanded energy and attention.

In short, it might be argued that the usual forces of insufficient knowledge, inertia and competing demands explain the implementation deficit: all that is needed is more time for the diffusion and institutionalisation of best practice. However, we suggest that the situation is more complex. If sustainable development genuinely offered tangible synergies between economic, environmental and social objectives, a persistent implementation deficit would be surprising. That such synergies remain elusive suggests that the 'deficit' cannot simply be read as a set of predictable obstacles to, and delays in, translating aspirations into practice. Instead, it points to a more fundamental dislocation between competing interpretations of what it means for development to be sustainable. And it is important to recognise that questions of process – how and by whom the debate should be conducted – have become as taxing, and as controversial, as those concerned primarily with outcomes.

Where next?

We have shown that an early and somewhat naive enthusiasm for sustainable development was replaced, during the decade after Brundtland, by more sober assessments of what could be achieved in practice. Attempts to turn the concept into 'real-world activity', as in land-use planning, placed the earlier broad (if not very deep) political support under considerable pressure. This was in part because of a latent recognition that commitments to sustainability might, at least in some instances, challenge the dominant conventional goals of market-led development and competitiveness. As its potentially radical implications became clearer, so conflict intensified over its meaning: all those involved in the planning process soon encountered the difference, outlined in Chapter 1, between the broad concept and specific conceptions of sustainable development. Assessing the sustainability of different patterns of land use turned out to be inseparable from judgements about their feasibility and desirability, and these were often profoundly in dispute. A system characterised by deeply ambivalent objectives, complex and extensive links in the policy chain, actors with some discretion, and susceptibility to outside interference is not conducive to clear-cut 'implementation' (Sabatier and Mazmanian 1979). Seen in this light, the implementation deficit that we have described is such a permanent feature that the phrase itself becomes a misleading description of the situation. Not only will policies be remade in the process of implementation but so too may the principles on which they have been legitimised. Acknowledging this much may not seem to take us very far, but it does at least move beyond simplistic notions of deficit, which imply some widely endorsed template for sustainability that can readily be applied in particular locations.

Significantly, however, rather than being consigned to oblivion once obstacles were encountered (as might be expected in an 'issue-attention' cycle), sustainable development has strengthened its rhetorical grip, while different interests have taken up arms about its meaning. Some groups have sought to maintain its radical edge, others to incorporate it seamlessly into the *status quo*, confining issues to safe areas of debate and accommodating conflict, in time-honoured manner, 'through the development of new institutions, compromise legislation and the appearance of acceptable explanations and solutions' (Sandbach 1980: 35). However, it would be difficult to deny that important changes have taken place. New ideas, as Marquand (1988: 12) suggests, have been 'thrashed out in the give and take of debate', and we might postulate that institutional learning has occurred within a framework of power relations, which are themselves not immutable.

The developments reviewed in this chapter suggest that planning has been an important forum for such learning, albeit an imperfect one. The nature of this process, the success (or otherwise) with which powerful interests have sought to capture the concept of sustainable development, and the significance of any changes on the ground are important questions for the remainder of this book. But before developing these ideas in the light of particular planning instruments, policies and conflicts, we must explore, in the next chapter, alternative conceptual bases for sustainable development in the context of land-use change.

3

INTERPRETING
SUSTAINABILITY

... when concepts (and hence the words that refer to them) become politicized, a struggle over meaning and morality takes place. When to classify is to decide, rival moral judgements contend for supremacy.

Wildavsky 1993: 47

The merit of any definition depends upon the soundness of the theory that results; by itself, a definition cannot settle any fundamental question.

Rawls 1972: 130

Introduction

We have argued that the concept of sustainable development became powerful at all levels, at least rhetorically, during the decade that followed publication of the Brundtland Report. Most striking, perhaps, were the high expectations for its reconciliatory potential, expectations that were particularly prominent in the context of land-use change. In practice, land-use planning – with its relatively visible procedures and tangible results – has proved to be one of the most important arenas in which conceptions of sustainable development are contested. Here more than anywhere else, it has become clear that trying to turn the broad consensual principle into policies, procedures and decisions tends not to resolve conflicts but to expose tensions inherent in the idea of sustainable development itself. One response to these difficulties has been to blame the lack of a coherent, *a priori* definition, and thus to seek to move beyond Brundtland towards rigorous and robust conceptions. Others have preferred to set aside the perplexing issue of definition and to concentrate on putting the broad idea of sustainable development into practice. Skolimovski (1995: 70), for example, is unrepentant about the use of a vague concept: 'It is better to muddle through to salvation than go crisply to damnation' (see also Blowers 1993b; Holdgate 1996).

In practice, conceptual and operational approaches proceed in parallel. Sustainable development cannot be theorised like gravitation, but, as with

28

other broadly consensual concepts such as liberty and justice, it has to be constituted in ways that might be defensible. All conceptions invoking institutions and decision rules will be normative and must co-evolve with practice in a wider political framework, which they might also challenge. The difficulty of operational definition can be seen to indicate not that a concept is without merit but that new ideas are being conceived and tested (Meine 1992). In this sense, sustainable development serves 'as a reminder of the chief concerns to which any policy must relate' (Banner 1999: 202). Different conceptions expose assumptions, beliefs and contradictions, and thus become a basis for debate, reflection and further refinement of principles and policy prescriptions.

This process of definition, challenge and reflection is well illustrated if we take, as a starting point, a set of interpretations of sustainable development that gained considerable currency in the post-Brundtland era: interpretations based on concepts of environmental, human and human-made capital. Our objective in this chapter is to show how elaboration of one general model can serve to expose core issues and divergent values, and to generate alternative conceptions of sustainability compatible with different interests and beliefs. We begin with 'environmental capital' because this approach, as well as having some general appeal for those seeking operational guidance, proved beguiling for many of those engaged in planning, where it became, on some accounts, 'widely used and influential' (CAG and Land Use Consultants 1997: 2). However, we shall show that what at first sight seems to provide a promising algorithm to guide land-use change reveals, on closer inspection, that decisions about what is 'sustainable' are inseparable from moral and political choices of the highest order. We argue that ethical dilemmas, judgements of value and important questions for liberal democracy must ultimately be confronted whatever our starting point, simply because they are so fundamental to the renegotiation of social and political priorities.

It might be objected that the quest for defensible interpretations of sustainable development is naive in its implicit assumption that outcomes are responsive to reason and principle, when there is ample evidence to suggest that policy and planning conflicts are actually resolved in favour of powerful vested interests (Blowers 1984; Flyvbjerg 1998; McAuslan 1979; Sandbach 1980; Wynne 1975). But recognising such structural constraints does not absolve us from the need to 'support our conclusions with viable principles' (Johnson 1991: 5). Argument is important in the policy process. Indeed, the exercise of power itself often requires a legitimising discourse, and some would see the discursive power of reason and persuasion as significant in challenging dominant interests and ideologies (see, for example, Hajer 1995; Litfin 1994; and, for a discussion, Radaelli 1995); at least we might agree with Forester (1994a: 180) that it is 'too simple and sweeping ... to say that practical judgement in planning simply works to support the powerful'. To show how the concept of sustainability has been interpreted, and different interpretations defended, is therefore not merely an exercise in semantics; it can help to expose the choices and judgements that determining what is sustainable must entail, and it may enable us to see more clearly the connections between new ideas, argument and established forms of domination and control. Where relevant, we relate the abstract frameworks discussed in this chapter to the real planning dilemmas on transport, nature conservation and minerals extraction that we analyse in more detail in later chapters of the book.

Environmental capital: a metaphor?

While Brundtland's consensual principle was widely endorsed, the quest for policy relevance was initially dominated, at least in the UK, by environmental economists (see, for example, Bateman 1991; Hanley *et al.* 1990; Pearce *et al.* 1989; for a later discussion, see Foster 1997). These analysts linked Brundtland's well-known axiom about the needs of present and future generations with the view that opportunities to generate well-being depend on the availability of various forms of capital. In this formulation, sustainable development implied that future generations should inherit no less opportunity for well-being than that enjoyed by the present incumbents, and the rule derived from such reasoning was that the productive potential of a stock of capital should at least be maintained over time. This was an idea with long antecedents (for example, Hartwick 1978; Hicks 1946; Locke 1988) and deep resonance in economic and managerial good sense.

If this is the starting point, it remains to establish an appropriate relationship (in terms of human well-being) between human-made and human capital on the one hand, conventionally including technology and infrastructure as well as knowledge and skills, and natural capital on the other, including resources such as fossil fuels and forests, and the functions of the biosphere (see, for example, Holland 1994; Pezzey 1992; Solow 1986; Spash and Clayton 1995; Victor 1991).[1] Many positions have emerged along a spectrum from 'weaker' to 'stronger' versions of sustainability. At the weaker end, while commentators accept that the positive (economic) value of environmental capital might usefully be given a higher profile in public policy, they remain sceptical that much in the environment has such special qualities that it should never be a candidate for substitution (see, for example, Solow 1992; for a vigorous debate, see Beckerman's 1994 broadside on sustainable development and the responses of Daly 1995; Jacobs 1995; Holland 1997). In stronger interpretations, reflecting diminished faith in infinite substitutability, positive rates of discount and long-term ecosystem resilience (Pearce and Turner 1990), environmental systems are seen to provide vital functions that could not plausibly be replaced by human action (see also Daly and Cobb 1989; Jacobs 1991; Spash and Clayton 1995; Turner

1988; Victor 1991).[2] Given imperfect knowledge of what it is that might be vital, and limited human capabilities, precaution then suggests the passing on of some essential natural resource base 'intact' (Pearce and Turner 1990: 238). Although it could be interpreted in very different ways, the metaphor of natural environments as 'capital' had broad appeal, extending well beyond an environmental constituency. The International Chamber of Commerce (1992: 1), for example, was content to acknowledge that 'sustainable development is about learning to value, maintain and develop our environmental asset so that we live off its income not its capital'.

But maintaining natural capital is clearly not compatible with unrestrained growth of the kind that places ever greater demands on resources and ecosystems (Boulding 1966; Pearce *et al.* 1989). Rather, it implies that 'we observe the "bounds" set by the functioning of the natural environment in its role of support system for the economy' (Pearce and Turner 1990: 52). However, where the 'bounds' lie, and their implications for growth and development, have been matters for intense debate, in which the scope for substituting one source of well-being for another has been a central issue. Weaker versions of sustainable development see the bounds as relatively unconstraining, whereas stronger interpretations can be regarded as environment-led.[3] Since even in the latter it is logically untenable to propose that *all* natural capital should be treated as inviolable, it has been necessary to make some distinctions. Thus the most vital elements of natural capital have been deemed 'critical', with the implication that their integrity should be protected in all but the most exceptional of circumstances. In other instances, it has been regarded as sufficient to maintain the overall quality or functioning capacity of the environment ('constant natural capital'). Damage caused by development could then be offset by compensatory measures, such as environmental restoration or habitat creation: 'investments to restore the natural capital of the supporting environments to some original position' (Pearce and Markandya 1988: viii). By combining stringent protection of what is critical with environmental compensation for losses or damage elsewhere, development could then be reconciled with the maintenance, or even enhancement, of a natural capital stock.[4]

However, these distinctions have been complicated by widespread agreement that aspects of the environment valued for spiritual, aesthetic or intrinsic reasons, and not just because they are essential in some material sense, could be located in the critical category (see, for example, Countryside Commission *et al.* 1993; Owens 1994; UK Government 1994a, 1994b, 1995b). Qualities that might justify this special treatment, such as uniqueness and irreplaceability, separate neither what is natural from what is human-made (Holland 1994),[5] nor what is essential to human life from what contributes to its quality and meaning: they can be found in semi-natural ecosystems as well as in cultural landscapes and monuments. Significantly, this extension shifts delineation of what is critical more obviously

31

and exclusively into the realm of values and judgement, and in doing so reopens an old divide. Just as the nineteenth- and early twentieth-century American conservationists of soils, water and forests found their doctrine of 'maximum sustainable yield' challenged by those for whom the environment embodied non-instrumental values (Hays 1987), so contemporary debates about criticality encompass much more than the material significance of environments. Expanding the category of what is critical beyond 'support systems for the economy' stretches the metaphor of 'capital' (with its human-instrumental connotations) but articulates a widely held view that sustainability should not be defined solely in terms of survival, health or opportunities for future economic development.

One can begin to see why maintaining natural or environmental capital (the term we shall now adopt) was attractive to many involved in planning, particularly those who felt that conventional approaches were resulting in an alarming (often cumulative) loss of much that mattered in the environment. It is not difficult to imagine that for a given geographical area – a region or a county, for example – 'environmental capacity' could be defined in terms of the need to protect critical environmental capital together with some version of the constant environmental assets rule (policies such as 'no net loss of woodland', for example). Indeed, a number of planning authorities in the UK attempted to do just this in the 1990s, using methods that we discuss in more detail in Chapter 4. For our present purposes, it is sufficient to note the theory that capacity would be breached if features or functions identified as critical were damaged or if other aspects of environmental quality were diminished in net terms. Development could be accommodated to the extent that it could reasonably be expected not to have such consequences. The amount permissible in any period would therefore depend on the potential – conceptual and technical – to eliminate, mitigate or compensate for environmental damage, and upon the degree of precaution considered appropriate in the face of uncertainty. Since sensitive development may avoid permanent loss and can sometimes result in environmental improvement, the relationship between capacity and growth need not be deterministic. Even so, on any typical planning timescale, a strong interpretation of sustainability must translate into limits to the amount of development (at least as conventionally defined in planning systems) that can be accommodated in a given area. If that development cannot be displaced, it might also imply ceilings on certain types of economic activity at national (or even international) level. Such interpretations of sustainability have clear resonance with earlier discourses on 'limits to growth'. It is easy to understand, therefore, why those with interests in economic development might prefer weaker interpretations of sustainability, emphasising integration, substitutability and faith in technical capabilities for ameliorating environmental impacts. But interests are not always easy to disentangle from values and beliefs (Sabatier and Jenkins-Smith 1993) and must usually be defended in some degree. Thus, in the process of making policies and decisions, divergent interpretations of sustainability draw upon and reflect not only a range of scientific judgements (themselves entangled with interests) but also conflicting moral and political doctrines. Age-old dilemmas, as we now go on to show, are never far below the surface in land-use planning conflicts and contribute much to their intensity.

Ethics

Perhaps the most immediate question arising from the framework discussed above is how to identify critical environmental capital. To say that some aspect of the environment should be placed in this category is (by definition) to claim that it ought to be protected intact, which in turn presupposes some rationale. When damage to environmental functions would pose severe threats to survival, human health or vital aspects of economic activity, an obvious rationale would be prudent self-interest: naturally, as Foster (1997: 233) points out, 'we want to avoid blundering into ecosystem catastrophe'. In such cases, weak sustainability might be sufficient: the most important functions of the environment could be deemed 'critical', not in the sense that they must be removed from the arena of trade-off but because any trade-off is so strongly in their favour, even in the short term. Application of the precautionary principle might move us towards the stronger end of the spectrum, placing some limits on short-term substitutability, but the case for protecting certain environmental assets intact still rests on the assumption that such a policy will optimise human welfare. It differs little from economists' traditional concerns with optimality, except in acknowledging that some environmental systems might constitute essential and non-substitutable material components of welfare, at least for the foreseeable future.[6] The precise location of any particular 'criticality boundary' (Blake 1999: 13–14; see also Jacobs 1991), reflected in environmental quality standards or measures to protect resources, will always be a matter of dispute, since thresholds 'can be informed by scientific understanding of nature's properties, but ... become determinants of decision making through political judgement and social choice' (Jacobs 1997b: 67).[7] The choice of protective measures is also likely to be contested. But the principle underlying prevailing definitions of what is critical in this domain is fairly clear: the environment is of instrumental value in human material welfare, and a requirement to protect it can be furnished by a utilitarian ethical framework in which the right course of action is that which tends to deliver the greatest well-being. As Norton (1982: 326) observes, 'if present behaviours soon lead to an interruption of production, then utilitarianism will be able to generate an argument against their conduct'.[8]

Things become more complicated when we turn to aesthetically pleasing, historic, tranquil or wild environments, which many would seek to include in the critical category, although they are less obviously crucial to material well-being. These aspects of environmental quality have featured prominently in land-use conflicts: for example, they were important in the planning dilemmas with which we opened this book. But as we noted above, it is particularly difficult in this domain to determine what might be vital and non-substitutable. Some environmental economists have suggested that weak sustainability would once again be sufficient to provide an adequate case for protection, because such environments make an important contribution to human welfare, this time in non-material terms. Thus (they would argue) the rationale for protection of these less tangible assets could also be provided within a utilitarian framework.

It is broadly within such a framework that concepts of environmental capital can become associated with the higher profile and controversial project of monetary valuation of environmental assets, and in particular with the inclusion of environmental considerations (both material and non-material) in cost–benefit analysis.[9] The argument is that if preferences for non-market assets (clean air, species, beautiful landscapes) could only be revealed or stated, they would turn out to be strongly positive and could often outweigh whatever preferences might be satisfied by development. This premise, and the multitude of valuation techniques that it has spawned, have been the subject of a wide-ranging, sometimes acrimonious, debate, which has been extensively covered elsewhere (for different perspectives, see Bowers 1997; Kelman 1990; J. O'Neill 1993, 1997; Pearce and Turner 1990; Pearce *et al.* 1989; Sagoff 1988; Spash 1998; Victor 1991). The problems are legion (and we revisit some of them in Chapter 4). Even in relatively straightforward material cases (as when a pollutant has an acknowledged serious impact on human health), the limits of science and of techniques for revealing preferences mean that claims about benefits and costs can always be (and nearly always are) contested. But a more fundamental difficulty for those seeking a generous interpretation of what is 'critical' is that when preferences are aggregated, the outcome might indicate that we should *not* save a landscape, or protect health, rather than (say) construct a road that would enable a large number of road users to save many small increments of time. As Onora O'Neill (1997: 131) argues, utilitarianism and environmental protection are 'uneasy allies' because:

> There is ... no guarantee that widely shared or trivial short-term pleasures that damage the environment will not outweigh the pains caused by that damage. The destruction of wilderness or environmentally sensitive areas will be a matter for concern only insofar as it is not outweighed by the pleasure of destroying them.

It is precisely because of such thinking that the metaphor of critical environmental capital was so appealing for conservationists, many of whom sought to retain the policy implications of criticality while detaching it from any economistic associations. Critical environmental capital seemed to provide a category in which, to borrow Hargrove's (1992: 161) words, certain things could be 'set aside' and exempted from use.[10] One would no longer have to demonstrate that the benefits of protecting some habitat or landscape outweighed the economic gains from development; it would be sufficient to point to its critical status. But assertion is clearly not enough: before such decision rules apply, particular environments would have to be classified as critical for some *a priori* reason – despite utility considerations rather than because of them – and in the process, as Wildavsky (1993: 47) predicts, 'rival moral judgements contend for supremacy'. To insist that protection is right *even if* the costs apparently outweigh the benefits is, in effect, to call upon frameworks that place 'limits on what would otherwise be the implications of aggregative [human] welfare calculations' (Keat 1997: 40).[11] This might point in the direction of deontological ethics, based on rights or duties, requiring that agents 'refrain from doing the sorts of things that are wrong even when they foresee that their refusal to do such things will clearly result in greater harm (or less good)' (Davis 1993: 206).[12] For some, it means expanding the moral community to embrace future humans, or the non-human world, or both, more fully and fundamentally than traditional frameworks, whether utilitarian or deontological, have allowed. Others find a rationale for protection in objective conceptions of what constitutes a good (human) life.

In exploring these alternative perspectives, we begin with the familiar refrain that environmental assets should be protected 'for future generations'. We deal with this first because, although often cited as self-evident, it is a claim that can hardly stand alone: it needs to be grounded within some wider ethical framework. The problem of inter-generational justice, however, 'subjects any ethical theory to severe if not impossible tests' (Rawls 1972: 284). Constructing a case for environmental protection around the interests or rights of future people is certainly less easy than the rhetoric might imply and has stimulated an intricate debate (for a range of views, see Laslett and Fishkin 1992). It is first necessary to establish that our descendants (even into the further future) have ethical standing, a task complicated by the fact that traditional concepts of benefit and harm 'are tied to notions of identity that give way when choices involving future people must be made' (*ibid.*: 1).[13] If, as some have argued, it is enough that our actions might impact upon specifiable others, even if they cannot be individuated,[14] the question remains as to what constitutes morally appropriate behaviour towards those who will exist in the future. Making reasonable assumptions about their basic needs might take us a good way towards protection of critical environmental assets with material and instrumental value; precaution would take us further. Even so, technological optimism and (within welfare economic frameworks) positive discount rates will pull in the opposite direction (Pearce and Markandya 1988; Hanley and Spash 1993).

Moving into the realm of less material and non-instrumental values (and further away temporally), we encounter more vexing problems in identifying the interests or preferences of future people, let alone in establishing rules for their consideration. As Sagoff (1988: 63) remarks, future generations may not miss what they have never had: perhaps a 'pack of yahoos will *like* a junkyard environment'. For some, therefore, a robust defence of environmental quality must lie in ideal regarding principles linking past, present and future. Thus 'our obligation to provide future individuals with an environment consistent with ideals we know to be good is an obligation not necessarily to those individuals but *to the ideals themselves*' (*ibid.*, emphasis added). In this view, we should strive to ensure that future generations 'belong to a community with ourselves', that they are capable, for example, of appreciating the goods of the non-human environment, and of contributing to them (J. O'Neill 1993: 34). If modern society produces 'mindless consumers', then the present generation fails not only the future but also itself and the past (*ibid.*). Although there is not scope here for a full discussion of environmental justice between generations, it is quickly apparent that appeals to futurity, although they have been so powerful in the discourse of sustainable development, often beg the question of 'what we know to be good', a difficult but central issue that we revisit below.

A case for 'setting aside' certain environmental assets, embracing considerations of futurity, might be made within a deontological framework by showing that there is an obligation to protect them, so that it would be morally wrong to do otherwise even if the consequences in terms of preference satisfaction would be better. Onora O'Neill (1996, 1997) attempts to do this in her constructive account of practical reasoning, in which, having shown that utilitarian thinking may not furnish adequate arguments for protecting environments, she suggests that a fundamental obligation 'to reject the principle of injury' might take us much further.[15] While such an obligation would not require agents to refrain from all injury, it would require them to refrain from systematic or gratuitous injury, including indirect effects arising from damage to natural or semi-natural environments:

> The basic thought ... is that it is wrong to destroy or damage the underlying reproductive and regenerative powers of the natural world because such damage may inflict systematic or gratuitous injury (which often cannot be foreseen with much accuracy or detail) on some or on many agents.
>
> O. O'Neill 1997: 137

Although, in Onora O'Neill's account, it is 'agents' (including distant others) who are the subject of fundamental obligations, and only humans are seen as having full capacity for agency, her argument is that the natural world would benefit indirectly because there would be a strong case for protecting the 'shared environment of human and non-human life' (*ibid.*: 137).[16] The fundamental obligation not to injure other agents gratuitously or systematically would be discharged in particular contexts through a range of positive and institutional obligations, for example to promote transport policies that do not contribute to dangerous climate change or agricultural practices that do not irreversibly damage biodiversity. Although the environment is treated as the 'material basis for human life and livelihood' (O. O'Neill 1996: 168), and 'injury' is quite tightly defined (to include damage to bodies and minds or coercion, for example, but not – or not obviously – deprivation of pleasing environments), this deontological framework could provide a generous rationale for critical status, going beyond what could be justified in utilitarian terms. It might, for example, require more stringent air-quality standards than principles of utility maximisation: it would simply be wrong to put human health at risk, even if the costs of reducing pollution to non-injurious levels outweighed quantifiable benefits in terms of illness avoided or lives saved.[17] It may be more difficult, on the grounds of rejecting systematic or gratuitous injury, to justify critical status for those aspects of the natural world that have (only) non-material value. For now, we note that the principle of precaution (as in utilitarian frameworks) would tend to justify wider boundaries for criticality. So too might the required virtues – imperfect obligations that are more selective and lack counterpart rights – which also form part of O'Neill's (*ibid.*) account. On this, she argues that an ethically sound relation to the environment must 'go beyond avoiding systematic and gratuitous damage that injures others' lives; it must also be expressed in care and concern to sustain and conserve at least some parts or aspects of that environment in a flourishing condition' (*ibid.*: 203).[18]

However, many commentators believe that a convincing case for protecting some environments, supporting what Norton (1982: 319) calls our 'intuitive ethic' for conservation, must ultimately depend on a less anthropocentric framework.[19] The quest for such a rationale goes some way towards explaining the enormous growth in literature on environmental ethics and the attempts by many writers to establish that non-humans have 'goods of their own', unconnected to any instrumental value to present or future human beings, and, furthermore, that this makes them directly morally considerable. Some then seek to include the pains and pleasures of non-humans in a utilitarian framework, others to confer upon at least some of them certain fundamental rights (Hargrove 1992 provides a useful review; see also Clark 1977; Johnson 1991; Leopold 1949; Regan 1981; Singer 1976; Taylor 1986; Wise 2000). Deeming particular species or habitats to be 'critical' could be a powerful expression of such moral standing. We explore these issues in greater depth in Chapter 6, where we are more directly concerned with the interface between land-use planning and nature conservation. The point to emphasise here is that some claims for environmental protection, or for critical status, depend upon beliefs (and increasingly draw upon carefully constructed arguments) about the moral status of the non-human world. The implications of these claims for policies and decisions are not always compatible with those arising in more anthropocentric frameworks. We also note that such biocentric thinking cannot, except by coincidence, justify critical status for important aspects of human and cultural environments.

This brings us to one further tradition, which is of particular interest because (when applied to environmental protection) it tends to blur anthropocentric and biocentric rationales and readily embraces the cultural as well as the natural. This is the view, advanced by John O'Neill (1993) among others, that an appreciation of the intrinsic worth of nature is one of the things that is constitutive of human flourishing; like friendship, it is part of what it means to lead a 'good (human) life' (see also Banner 1999; Goodin 1992; Hargrove 1992; Partridge 1984). Within this framework, the goods of non-humans ought (for the most part) to be promoted. For John O'Neill (1993, 1998), this is part of a larger argument in which he seeks to defend objective conceptions of the good,[20] an argument that has wider implications for the role of planning in making difficult choices.

It is useful to draw together the arguments at this point. Some aspects of the environment could be deemed critical – not to be damaged or destroyed – whether one adopts a utilitarian or non-utilitarian framework, an anthropocentric or more biocentric one. In many cases, however, claims and counter-claims about criticality must depend not only upon different degrees of precaution but also upon different views (not always self-consciously held) about what is good and what is right, and about the boundaries of moral considerability.[21] The positions taken up in planning conflicts can often be attributed to the identifiable interests of protagonists; even so, they normally have to be defended, and it is in this process that the various rationales will be employed, exposed or appropriated. What might have seemed to be a technical problem of classifying environmental capital (and planning accordingly) exposes enduring and fundamental questions; sustainable development is then interpreted on the basis of different interests, values and beliefs. And while we have focused this discussion on the determination of what is critical, the issues are scarcely less challenging in the context of non-critical environmental capital. It will still be necessary to decide how much it is right to protect overall and what might constitute appropriate compensation in the face of unavoidable damage or loss. The potential for restoration, relocation or recreation of habitats or landscapes depends not only on technical capabilities but also on what the original was valued for, and why. Thus, for example, the impressive skills of mining companies in restoring the agricultural capacity of land fail to placate those concerned about irreparable loss of meanings attached to landscape, or those for whom 'naturalness' or 'wildness' provides a vital reflective counterpoint to the human world (Elliot 1997; Goodin 1992; Holland 1997).[22] The important point here is that since the underlying ethical questions are not of the kind that can quickly (or ever) be resolved, defining what is 'sustainable' was never likely to be an exercise in consensus. The best we might hope for is that alternative moral frameworks will point happily in the same policy direction, at least in particular cases. There may indeed be some – even much – common ground (Beatley 1994; Turner 1988), but, as forthcoming chapters on transport, nature conservation and minerals extraction will show, many instances remain where divergent conceptions of sustainability indicate fundamentally different choices about the use and development of land.

Politics

So far, we have concentrated on different rationales for environmental protection and have dealt only imprecisely with the benefits of development – benefits, in the terminology of the environmental capital model, that flow from what is human-made. Yet the extent to which proposals for physical change, addressed through the planning system, contribute to development in the sense of social progress is equally fundamental to different conceptions of sustainability. This points to an important area where conceptions based on the metaphor of capital came to be seen as insufficient. In focusing on flows of benefits from different forms of capital over time, these models tended to treat generations as single, global meta-individuals (Hanley and Spash 1993) and had little to say explicitly about *intra*-generational justice and distribution (for further discussion, see Harvey 1996).[23] This was changing by the late 1990s, when social considerations were explicitly added as a third dimension to conceptions of sustainable development in the UK, although usually without progressing towards the kind of decision rules that the environmental capital model (however inadequately) had tried to develop. This is important, because repeated assertion that environmental protection, economic development and social justice are mutually interdependent (or at least are not invariably at odds with one another) cannot disguise the evidence that these objectives are certainly not always harmonious (Dobson 1998). In the world of policies and decisions, legitimate interests, rights and obligations frequently conflict, and some claims are overridden, for good reasons or bad, as values come into play with the exercise of power. Foster (1997: 235) adroitly sums up the implications: 'Values, as it were, can be left to disport themselves in happy plurality when off-duty, but the need for a decision calls at least the relevant ones rather sharply into line'.

This 'calling into line' is particularly sharp when land-use policies have to be adopted and decisions made. In any given area, planning might require a defensible definition of environmental capacity along the lines that we have outlined above. As well as working within national parameters (such as air-quality standards or site designations), planners might wish to extend protection to locally important cultural landscapes or habitats, preserve tranquillity in particular places, or nurture a sense of history and civic space in towns. But most areas will also be subject to development pressures related to a range of anticipated demands, whether these are manifest in markets or through the provisions of the state, or via some combination of the two. Familiar examples include requirements for industrial and commercial development, transport infrastructure, housing and minerals extraction. Respecting environmental capacity may mean that not all such demands can be accommodated. But since determining what is critical, or what should be held 'constant', involves arguments about needs, preferences and values, it will rarely – perhaps never – be possible to establish capacity in isolation from consideration of conflicting priorities. The questions of how categories of critical and constant environmental capital might be defined and defended now become entangled with others concerning the definition, articulation and status of various demands and their relationship with social needs. It is here, and not only in refining methods of delineation, that more work needs to be done.

Suppose, to take an example that we develop in Chapter 7, that the demand to extract minerals in a particular area is irreconcilable with environmental capacity as the relevant planning authority has sought to define it,[24] and that there are no unexploited technical or spatial fixes that would permit conflicting objectives to be achieved simultaneously. Decision makers are then faced with at least two sets of claims: on the one hand, that minerals are essential to the economy and to society and must be provided, albeit with 'least pain'; on the other hand, that certain environments in this area are not for trading off, and that loss of others must be compensated for in appropriate ways. They are also likely to be influenced by the 'magic symbols' of jobs and prosperity (Caldwell and Shrader-Frechette 1993: 33), with implicit or explicit implications for social justice. A familiar argument, in which the pro-development case is defended on grounds of preference satisfaction, would be that demands for minerals should be met, that this would improve well-being, and that the welfare gains would outweigh environmental losses. Losses should of course be minimised, but weak sustainability fails to furnish a case for protection, and the claims of environmental capacity must be loosened (at the stage of making plans) or overridden (in development control). If these are the arguments for more rather than less development, defenders of the environment (seeking to define sustainability in 'strong' terms) might rescue the case for conservation by adopting ethical frameworks that are not grounded in preference utilitarianism, or that are less anthropocentric, or both. They might argue that preferences cannot carry moral weight if satisfying them means breaching rights or obligations, thus violating the principles of justice (Rawls 1972). Alternatively, they might try to show that 'the wants which people happen to have' (Barry 1990: 38) – for more roads, for example – are simply not the 'right' ones, and that to satisfy them would not in fact improve well-being.[25] Whether implicitly or explicitly, those seeking to establish a robust case for environmental protection have often drawn upon such frameworks, in effect suggesting that the preferences to be satisfied by development should have no prior claim on public policy. Policy should not be based on the aggregation of preferences that people 'happen to have' (J. O'Neill 1993, 1998; Sagoff 1981, 1988).[26]

However, the case for conservation cannot be rested there, as is sometimes implied. Few ethical frameworks (except perhaps the most biocentric) are likely to result in an automatic warrant for protection. For example, one could imagine circumstances in which an obligation to 'reject the principle of injury' would be breached if development does *not* proceed. After all (to pursue our hypothetical case), the working of minerals provides employment and income, and its products construct homes, schools, hospitals and public buildings, helping to satisfy not just 'trivial pleasures' but also basic needs; the latter might justly exert strong claims on public policy. Or, if drawn to objective accounts of the good, one might argue that efforts to ensure that people are creatively employed, housed or educated could hardly be construed as misguided. Although the desirability of meeting demands (without judging their merits) *is* frequently offered as a rationale, the case for development *need* not rest upon maximising the satisfaction of want-regarding preferences. Certain human requirements can be regarded, just as much as critical environmental capital, as special and non-negotiable,[27] and developers too might legitimately have recourse to the claims of social justice, rights (including property rights) and obligations, or might invoke objective conceptions of well-being.[28] And such anthropocentric considerations remain significant, even in ethical frameworks that accord a prominent position to the interests or rights of non-humans; although the relevant moral considerations might differ (as might outcomes), the fundamental problem of reconciling the legitimate claims of humans and non-humans remains (Beatley 1989; Norton 1982; Regan 1981; Taylor 1986).

In short, there are no algorithmic answers to the question of 'what is sustainable?' Persistent misgivings about utilitarianism, and about the appropriateness (at least for some goods) of trade-off in markets or surrogate markets, points to frameworks in which we might be guided by different ends or principles. But none provides 'an auto-pilot for life' (O. O'Neill 1996: 78); all underdetermine action and leave substantial scope for judgement. While meeting multiple requirements is central to the concept of sustainability, in developing meaningful conceptions we cannot escape the need to adjudicate between claims that all carry moral weight (Holland 1997: 129).[29] What seems to be required is some way of deciding what is of fundamental importance in environmental, economic and social domains, allowing for trade-off where appropriate but without allowing that everything can be traded off against everything else. In difficult cases, we have to be able to say 'this is more important than that', or 'this is more valuable than that' (Attfield and Dell 1996: 43), allowing for some goods (or obligations, or rights) to take precedence over lesser considerations. Immediate threats to human life, for example, might claim priority over protection of significant habitats, but demands for new roads or retail facilities might not. Some will immediately detect in such stratification a challenge to (modern) liberal neutrality between different conceptions of the good (Wissenburg 1998), and it is clearly quite incompatible with postmodern relativism.[30] But developing any meaningful conception of sustainability is bound to demand judgements about what is good and right, even when proponents do not self-consciously engage in an exercise of moral and political philosophy. We have certainly travelled a long way from any notion that delineating different forms of environmental capital could be a straightforward technical process, or that it might point clearly (in the context of land-use planning) to decision rules for development. Little surprise, then, that sustainable development has failed to provide the resolution that many had hoped for and expected.

Nor is the exposure of these questions an artefact of our particular point of departure: it seems likely that, wherever we start, determining what is sustainable takes us inexorably towards certain fundamental choices. Two examples, one moving in the direction of more specificity and one towards greater generality, will suffice to illustrate this process. The first is a refinement developed by the English statutory agencies, which were finding environmental capital a 'surprisingly difficult' concept to apply (CAG and Land Use Consultants 1997: 86). Focusing on environmental services rather than on specific entities, the agencies have sought a 'consistent, systematic and transparent framework' with 'a finer grade of management implications' than they could derive from the standard categories (critical, constant and tradable) of environmental capital (Cole 1997: 4, 5). But it remains unclear how even the most subtle delineation of what is important and irreplaceable could avoid the difficult issues outlined above, still less how a potentially reductionist methodology could constitute a 'considerable step forward' in settling questions conclusively (CAG and Land Use Consultants 1997: 86). The second (and politically more significant) illustration is what we might call the 'Panglossian' approach, incorporating economic, environmental and social considerations in which, through some magical process of integration, 'achieving all these objectives at the same time is what sustainable development is about' (DETR 1998d: 3).[31] But recognising the significance of all three dimensions cannot evade the fact that 'moral conflict is a real enough phenomenon' (Holland 1997: 129). As Isaiah Berlin (1969: 167) famously argued, 'not all good things are compatible, still less all the ideals of mankind'. Dilemmas arise (as we have shown above) when there are tensions between competing claims, and integration (or reconciliation) will not be achieved by *fiat*. No conception of sustainable development can be complete without some guidance on adjudication in such circumstances; and that has not been a prominent feature of definitions that set out a menu of potentially conflicting objectives.

Planning as a dialogical forum

One way forward, for many commentators, lies in dialogical processes through which it might be possible to explore values and differences, test the possibility of consensus and (where appropriate) arrive at defensible, inter-subjective positions (J. O'Neill 1993, 1998; Sagoff 1988; Wiggins 1999). Clearly not every policy or decision could be made in this way (the result would be paralysis), but there is a growing consensus that difficult interpretations and choices should be made through 'practical reason, exercised in judgement' (Banner 1999: 200) and grounded in empirically verifiable evidence, when it is available. This implies the need for a forum, a 'social, dynamic, creative and interactive context [for] the deliberative process' (Holland 1998: 16), and it is arguable that land-use planning might provide at least one such context. This much has been recognised, as we noted in Chapter 1, by a number of prominent planning theorists, and the idea has also found favour among practitioners (see, for example, TCPA 1999). However, most agree that there is a great deal of work to be done. Current planning systems fall notoriously short of any such ideal, deliberative approaches may be as vulnerable as any other to control and manipulation, and the boundaries between deliberative forums and more formal aspects of the democratic process are often left obscure. There are also substantial, unresolved problems of 'discursive competence' (Foster 1997: 244), involving questions about the appropriate role for expert and lay knowledge of various kinds.

If planning is to provide a forum for deliberation about sustainable development, a number of other difficulties must be confronted. The first is that planning authorities have generally been expected to judge not the merits of a proposed development (whether, in itself, it serves purposes that are good or right) but its acceptability in a particular location and the potential to minimise any negative impacts. In the UK, for example, applicants for planning permission have not normally been required to demonstrate the 'need' for a development.[32] Although it is important to show that proposals are in accordance with the development plan, such plans are often themselves required to make land allocations to meet various demands; in effect, whole classes of goods and services are deemed to be 'essential'. Although in some cases policy is informed by a concept of social need, in many others the role of planning may be better understood as one of promoting the efficient satisfaction of those 'want-regarding preferences' that people happen to have. Where considerations of 'need' have crept in, this has been contentious. A requirement to demonstrate 'need' for retail and leisure facilities, for example, became the subject of legal challenge, despite the fact that 'need' was interpreted primarily in terms of market demand and alternative provision, rather than in any more fundamental sense.[33] And when any proposal is subject to environmental assessment, so that it is necessary to show some consideration of alternatives, local planning authorities are often neither capable nor empowered to consider alternative technologies or policy approaches, rather than simply alternative locations (Glasson *et al.* 1994; Jewell 1995). This traditional remit leaves little scope for making distinctions between wants and needs or for considering whether and in what circumstances different kinds of preference might claim priority; it affords a somewhat cramped opportunity for dialogue about what constitutes sustainable development.

A second difficulty relates to the important spatial dimension in many planning issues. Conflicting interests at different scales and in different locations have long presented planning with some of its most intractable dilemmas, to which sustainability adds a new dimension. It is difficult enough to hold meaningful dialogue about the complex issues involved even in one self-contained area; all the more so when the dialogue must take place across administrative boundaries and different levels of governance. Take, for example, the task of maintaining environmental capital at a regional scale, along the lines that we have outlined in this chapter. This offers considerable scope for directing damaging development to 'suitable locations' (Cowell and Owens 1998: 797); indeed, because areas vary in their sensitivity, locational policies might allow for maximum development within any given environmental capacity and thus help to integrate environmental and economic goals. In practice, as we show in Chapter 7, such a strategy tends to concentrate locally unwanted land uses into locations already adversely affected, sometimes by exporting them altogether to more peripheral regions. Both the exporters seeking to protect environmental capital and the recipients giving high priority to the social objective of employment might see this solution as sustainable in a local context, but unless trends in production and consumption can be challenged, the effect might be to exacerbate spatial inequality on a wider scale (Blowers and Leroy 1994; Cowell 2000a; Cowell and Owens 1998). The more general difficulty is how to allow for the particularities of places and communities – acknowledging that 'sustainable development issues are not the same everywhere' (DETR 1998d: 24) – without abandoning all prospect of universal conceptions of sustainability.

One approach, for which there is a degree of precedence in planning systems (see Chapter 6), is a hierarchical one in which internationally important agendas (for development or for conservation) carry greater weight than matters of national significance, which in turn exert authority over what can be conceived of as purely local interests. Thus a European transport project might take precedence over a nationally designated wildlife site, but local employment considerations might defer to protection of an internationally important habitat. Even if it was always honoured, however, such an ordering is unsatisfactory because in focusing on the *scale* at which particular options are deemed significant, it deflects debate about the nature of the preferences, needs or values to be satisfied in each case. Such hierarchies of scale interact in significant, and as yet largely unexplored, ways with the ordering of moral significance that we discussed earlier in this chapter. The challenge for any dialogical institution is clearly very considerable.

A third set of difficulties relates to the role – and perceived role – of professional planners. With some notable exceptions, the profession has leaned towards an image of neutrality in which planners offer impartial advice to elected representatives on the formulation and implementation of policies (Tait and Campbell 2000; Thomas 1994); despite continued academic interest in 'advocacy planning' (Davidoff 1965; Forester 1994b; Heskin 1980), identification with particular causes or values has come to be regarded as 'unseemly' (Campbell and Marshall 1999: 475). This is not the place for an extended discussion of the 'pivotal' but ambiguous relationship between professionals and policy makers (Tait and Campbell 2000: 490), or about the credibility of the image of neutrality, but two questions are important in the context of our present discussion. One is whether planners would welcome the opportunity to provide an 'interactive context' for deliberation, and indeed whether they are equipped to do so (Tewdwr-Jones 2000). The other – if they do embrace such a role – is whether they should seek to act as neutral facilitators, expert participants or advocates of particular conceptions of sustainable development. Some commentators have argued that planners should be prepared to defend explicitly particular ethical positions on issues of social and environmental justice (see, for example, Campbell and Marshall 1999); in the context of sustainability, as we have seen, taking up virtually any position involves important ethical and political choices. If the planning system is indeed to develop any potential as a deliberative forum, then the relationship between elected representatives, professional planners and the wider public in this context will merit more rigorous attention.

Conclusions

We began this chapter by outlining an interpretation of sustainable development based on concepts of environmental, human and human-made capital, one that seemed to have particular resonance with, and influence on, the field of land-use planning. We showed that attempts to operationalise this framework – determining, for example, what constituted critical environmental capital, or environmental capacities – led inexorably towards important questions of value: what contributes to well-being; whose good is to be taken into account; whether maximising the good necessarily defines the right course of action; and where the locus for such important decisions should lie. Fundamental disagreement about such issues, many of which are the subject of ancient and enduring debate, becomes internalised into conflict over the meaning of sustainable development itself. It should not be surprising, therefore, that concepts of environmental capital have proved difficult to apply in practice and have failed to provide a straightforward means of resolving land-use conflicts. What *is* surprising is that anyone ever imagined that they would.

These conclusions do not seem to be dependent on our particular point of departure. All conceptions of sustainable development must involve ethical and political choices grounded in distinctive traditions. Even when such positions are not self-consciously held or articulated by proponents, and even when they may not be what determines outcomes, they underlie deep differences of view about what ought to be done. This explains why planning policies with divergent outcomes can be claimed by various groups to be 'sustainable'. However, we do not suggest that the failure to converge upon a 'figure of resolution' means that attempts to apply concepts of sustainable development to land use have been unhelpful or even futile. For one thing, they have probably invigorated the search for solutions that meet multiple objectives. More significantly, where such solutions are not available, the debate has performed a valuable function in exposing different positions and in clarifying the sorts of choices that deciding what is sustainable must entail. Failure to recognise the fundamental issues and challenges has all too often fuelled unrealistic expectations, or fatalistic acceptance of a 'Humpty Dumpty' view of sustainability.[34] Accepting such challenges, on the other hand, may allow for progress in practical reasoning to be made.

One implication of our discussion is that to make such progress, there is a need for dialogue, deliberation and the exercise of good judgement. The 'market' alternative – involving the aggregation and monetisation of individual preferences – seems ethically and practically inadequate for addressing the kind of land-use issues with which we are concerned. Policies for land must engage with scientific and institutional complexity, and typically involve a plurality of values, often held 'with deep emotion' (Caldwell and Shrader-Frechette 1993: 3). While systematic treatment is useful, wholesale reductionism is not. The crucial question then becomes one of how and where such deliberation should be conducted. We have asked whether the planning process itself might provide a dialogical forum for debate and judgement that could inform the difficult choices to be made. We suggested that it might be promising in this respect, and indeed if 'planning for sustainable development' is to acquire any real meaning, there must be some provision for deliberation within the system. But we also identify a number of potential difficulties. Some relate to institutional and legal structures: scrutiny of demand, for example, and therefore of some of the competing claims in land-use conflicts, has often been beyond the reach of land-use planning systems. Other difficulties relate to unresolved issues about deliberation: who, with what credentials, should 'deliberate'? How successfully, if at all, could such dialogue transcend boundaries and different levels of governance? What would be the (new?) relationship between participants, professionals and politicians?

The challenges of interpreting sustainable development highlighted in this chapter will be illustrated when we look in more detail at nature conservation, promotion of integrated transport and minerals policy. These more specific contexts also provide insights into the actual and potential role of the planning system as a forum for argument and debate. Before moving on to this detailed analysis, however, we have one more task to complete in establishing the broad framework within which planning systems engage with sustainability. We examine, in Chapter 4, a range of techniques and procedures that have been widely held to promote more sustainable outcomes in land-use planning. This enables us to scrutinise claims about objectivity in the planning process more closely and to develop further some of our arguments about the role of deliberation.

4

DEFINING AND DEFENDING

Approaches to planning for sustainability

> Until more sophisticated techniques of measurement and eval-
> uation are available there will continue to be doubts as to the
> durability of the principles of sustainability.
>
> Winter 1994: 890

> ... reasoning quickly turns into rationalization and ...
> dialogue becomes persuasive rhetoric under the pressures of
> reality.
>
> Flyvbjerg 1998: 5

Introduction

The search for rational ways of making land-use decisions – traditionally
afforded an important status in the planning process (Breheny and Hooper
1985; Thomas 1994) – has interacted with the rising profile of sustainability in interesting and significant ways. In a message that resonated with
planning policy communities, both the Brundtland Report and Agenda 21
emphasised 'assessing performance' in relation to sustainable development
(Hodge and Hardi 1997: 1). Governments and public agencies have shown
keen interest in techniques for measuring sustainability, evaluating policies
and monitoring progress; many academics have responded with enthusiasm.
The result has been a plethora of methodologies, techniques and procedures,
often conceived of as a 'toolbox' for implementing sustainable development.
These include new approaches as well as developments of established ones –
assessment techniques, indicators, audits of various kinds, 'footprint' studies,
ecological accounts – all claiming to provide better information or struc-
tured evaluation to promote more sustainable policy choices. Not everyone
is impressed. Some critics fear that the essentially political challenges of
sustainability are disguised by such procedures as technical matters of better
management and control. They argue that rather than perpetuating 'the
worthy but increasingly unrewarding search for "objective" methodologies',
we should be seeking 'institutional reform aimed at enriching and refining
open political debate about public values' (Grove-White 1997: 30; see also

Jacobs 1997a). In practice, both technical rationality and 'deliberative and inclusive processes' have become growth industries, so that experimentation with a variety of novel participatory forums has proceeded in parallel with the development of an 'audit culture' (Grove-White 1997: 22). We return later to the question of whether deliberation must necessarily be inclusive; for now, we note that such forums have often been brought together around statutory planning processes or non-statutory initiatives such as biodiversity action plans and Local Agenda 21.[1] None of this may be sufficient, however. For many commentators, a further prerequisite of sustainable development is 'integration', involving changes to the ways in which policies are made and institutions operate. Integration is seen as a way of ensuring that environmental, social and economic objectives are all given due consideration, sometimes within a specific spatial framework. While this may require structural change, it is often assumed to be aided by formal assessment techniques or by deliberative participation.

All of these approaches have been viewed in terms of 'operationalising' sustainability – translating a concept that is presumed to be agreed in principle into something workable on the ground. In this chapter, we adopt a more critical perspective, exploring the nature of the relationship between techniques and procedural reforms on the one hand and ideas about sustainable development on the other. We argue that appeals to 'better (usually better quantified) information', 'greater participation' or 'more integration' must be understood within a wider political context in which there are contested interpretations of sustainability and unequal relations of power. This means that assessing the *effectiveness* of different approaches – whether or not they lead to 'more sustainable' development – is far from straightforward and indeed may not be a fruitful exercise. Rather than accepting the analogy of a neutral 'toolbox', we consider how different forms of rationality become bound up with alternative conceptions of sustainability, and how they are deployed in the politics of land-use change. Increasing recourse to technical rationality, for example, might be seen as part of an 'analytical arms race', in which intangible environmental qualities must be rendered material in the planning process to match the apparently 'objective' status of other claims (Owens 1997; Owens and Cowell 1996). Deliberative rationality could be seen as empowering for participants, or as just another means of legitimation; in either case, new procedures for public involvement direct renewed attention to the relationship between process and outcome in the making of policies and decisions. And while integration in its various forms seems intuitively plausible as a precondition for sustainable development, there is a need for critical examination of the meaning and application of the term.

We address such issues by considering how technical, deliberative and integrative approaches have been deployed in planning for sustainability, and with what results. Our analysis is necessarily selective, and we could have selected differently, but in covering three broad sets of techniques and procedures and relating them to conceptions of sustainability, we hope to address the critical aims of this chapter. We look first at examples of appraisal or audit-type approaches, which might be regarded as prime ammunition in the analytical arms race, focusing on environmental assessment, capacity studies and monetary valuation. We then examine procedures that seek greater deliberation and more inclusive involvement in planning, together with some of the fundamental questions that they raise. Third, we turn to the quest for 'integration', paying particular attention to experiments with holistic environmental planning in the Netherlands and to prospects for achieving multiple objectives under the auspices of development plans in the UK.

An analytical arms race

Environmental assessment

We begin our discussion along the well-travelled path of environmental assessment, defined by Kennedy (1988) as both a science – involving methods for identifying, predicting and evaluating the impacts of particular actions – and a set of procedures for ensuring that analysis takes place and informs the decision-making process. With a long-established role in land-use policy, and a claim to mitigate the social costs of development, environmental assessment was bound to be recruited to the cause of sustainability. Proponents have seen it as providing vital information that leads towards better and more sustainable decisions, as if the facts will 'speak for themselves' (Emmelin 1998: 142). Much analysis and comment has dwelt on procedural aspects and on the quality of environmental statements, generating numerous proposals for refinement and methodological improvement (see, for example, CEC 1996; DETR and EFTEC 1998; Foresight Energy and Natural Environment Panel 2000; Goodland and Mercier 1999; IEEP 1999; Therivel et al. 1992). For some critics, however, assessment techniques epitomise the managerialist practices through which modern states legitimise dominant patterns of growth and development (Paehlke and Torgerson 1990; Wynne 1975); in this view, the quest for 'better' methods and procedures entirely misses the point.

In practice, environmental assessment has been neither straightforwardly instrumental (in the sense of promoting sustainable development) nor simply an exercise in technical rationality. Assessment of projects (originally, 'environmental impact assessment') has undoubtedly helped to identify and mitigate potential impacts, even if it has been unable to make 'precise, verifiable predictions' (Emmelin 1998: 138). Often, however, it has been little more than a compliance exercise, demonstrating that projects will satisfy existing environmental regulations. The presumption is that development, modified if necessary, should be able to proceed, and abandonment of proposals has been unusual. It might seem, therefore, that project-based assessment can mobilise only those conceptions of sustainability already embodied in development strategies and legislation (Sadler 1996), and that if these are weak it can lead at best towards an incremental greening of growth. But this would be to take too static a view of a longer-term process of policy change. Certainly, such a conclusion would overlook the opportunities for critical scrutiny that even a relatively restricted assessment process offers, especially as experience grows over time. Whatever its shortcomings, environmental assessment partially shifts the burden of proof, uncovers information that might not otherwise enter the decision-making process and exposes this information to a range of interested groups, even if not as early in the proceedings or as fully as some critics would like (Bartlett and Kurian 1999).[2] The controversial Lingerbay superquarry proposal, which we discuss in Chapter 7, provides a clear case in which environmental assessment lubricated the production of information and stimulated intense debate about the acceptability of ecological risks and social change.

One could also argue that repeated application has stimulated a process of policy learning, in which the inadequacies of project-based assessment have been exposed and ways opened through which stronger conceptions of sustainability might be mobilised. Opponents of specific developments have been frustrated by the failure of the system to consider genuine alternatives and by their inability to question the wider strategies of which individual projects form a part. Assessment on a case by case basis cannot always prevent environmental constraints being breached as the impacts of individual projects accumulate, and it sits uneasily with environment-led conceptions of sustainability. Acknowledgement of these problems explains growing support for strategic environmental assessment (SEA), in which the focus is expanded to plans, programmes and policies (Partidario and Clark 2000; Therivel et al. 1992). SEA is clearly more challenging than project-based assessment, not only because of the added complexity of predicting and evaluating impacts at a higher level but also because of the potential for raising searching questions about policies and development strategies, directing the 'light of examination' (Wynne 1975: 125) towards dominant institutions and values. This is particularly the case given the openness normally envisaged by its proponents, with public participation and reporting being regarded as 'basic elements in effective SEA' (Partidario and Clark 2000: 8).

This potentially radical challenge creates its own problems. Concern that SEA might be difficult to control probably accounts for the slow progress in establishing the necessary legislation and the tendency for strategic assessment procedures to pull in a technocratic rather than a participatory direction (Bina 2000; Bina and Vingoe 2000). A European Directive on SEA, first mooted in the 1970s, had still not been adopted by the end of the millennium, although its scope had been the subject of protracted negotiations (CEC 1997, 1999b).[3] The final version does not extend to policies, and assessment might have been further restricted to plans and programmes forming part of an official land-use decision-making process had not the European Parliament (2000) challenged this narrow remit. Although progress in advance of the Directive has been somewhat uneven, a number of member states and the European Commission itself have adopted SEA-type procedures, with some positive effects, particularly in terms of learning and communication.[4] In the UK, where the system is non-statutory, appraisal has become more strategic in certain policy areas (including transport and water resource planning), but elsewhere the application of SEA to sectoral plans and programmes has been 'relatively limited and *ad hoc*' (DETR 1998e: para 3.24; Therivel and Partidario 1996), and strategic assessment has had little apparent impact on many important national policies and programmes.[5]

At the local level, planning authorities have been expected to carry out environmental appraisal of development plans since the early 1990s (DoE 1993a), and this practice has spread and evolved, although, with some honourable exceptions, it has often been undertaken as an internal exercise with little public input (Therivel 1995). By the end of the decade, the government was urging that environmental assessment be absorbed into wider-ranging 'sustainability appraisals' of all plans (including regional planning guidance) in which 'the same methodologies … can be developed to encompass economic and social issues' (DETR 1999b: para 4.14; DETR 1999a, 2000c). This extension reflects a more general trend towards seeing strategic appraisal (like sustainable development itself) as a process that should encompass economic and social as well as environmental dimensions (Bina 2000). Such a move may seem both desirable and uncontroversial, but, as with the shift to 'Panglossian' interpretations of sustainability discussed in Chapter 3, it might also be read as a reassertion of economic primacy, in a pre-emptive challenge to environment-led interpretations of sustainability.

How, then, might we characterise the relationship between environmental assessment and sustainable development? To the extent that it raises the profile of environmental considerations and encourages greener growth, there is evidence that assessment makes development 'more sustainable', at least in the weaker sense. More significantly, perhaps, in as much as it provides for the more open scrutiny of proposals, environmental assessment has the potential to stimulate learning within and between different groups and to provoke deliberation about what sustainable development should mean. Indeed, these might be among the more useful effects of assessment procedures. Rarely, however, has this deliberative or critical framework been highly developed, either in the appraisal of individual applications or plans, or in the institutional arrangements that govern assessment and planning systems alike. SEA has particularly interesting implications in this context. In exposing the environmental consequences of wider policies and plans, it might aid those who favour stronger conceptions of sustainability, but for this very reason it is more obviously challenging to the *status quo*. Little surprise, then, that governments have approached strategic assessment with caution, or that they have moved to extend its coverage to economic and social issues, potentially neutralising any environmental threat to dominant modes of development.

An important implication of this discussion is that environmental assessment can mobilise different conceptions of sustainability, depending upon the context within which it is applied and what precisely is asked of the technique. The question 'what are the likely impacts of this project, and how can they be mitigated at acceptable cost?' is framed quite differently from one that asks 'what are the environmental implications of alternative ways of meeting this (social or economic) policy objective, and are these implications acceptable?' In the second context, environmental assessment, at the strategic level in particular, could assist in defining environmental capacities. At least it is hard to see how judgements about capacity could be arrived at without an assessment of sorts, or how this process could promote stronger conceptions of sustainability unless serious and irreversible impacts could be deemed unacceptable.

Assessing capacities for development

This brings us back to the concept of environmental capacity, which has enjoyed popularity among some planning authorities in the UK and, initially at least, seemed acceptable to government: the 1994 sustainable development strategy saw it as a function of the planning system to 'ensure that development needed to help the economy grow ... takes place in a way that respects environmental capacity constraints' (UK Government 1994a: para 35.4). One widely accepted application has been in considering the potential for towns and cities to accommodate further development on brown field land without damage to the quality of the urban environment (Arup Economics and Planning 1995; Counsell 1999c; DETR 2000b; Government Office for the East of England 2000). Another – more controversial – has involved attempts by a small number of planning authorities to assess the capacity of whole areas (usually counties) to sustain development. This they have done in much the way that we envisaged in Chapter 3, by considering the extent, value and replaceability of environmental resources (usually employing adaptations of traditional constraints mapping) and the probable impact of development upon them (see, for example, Babtie Group Ltd 1993; West Sussex County Council 1996; and for discussion, Counsell 1999c; CAG and Land Use Consultants 1997; DETR 1997a; Jacobs 1997b). Related conceptually, if not always in practice, have been techniques for delineating environmental capital, moving from classifications as 'critical', 'constant' or 'tradable' to more nuanced methodologies such as that developed by the statutory agencies and piloted by a number of local authorities at the end of the 1990s (CAG and Land Use Consultants 1999). Interestingly, the greatest benefit reported by participants in the pilot studies was the potential to make the reasoning behind policies 'rigorous' and 'inquiry-proof' (Levett 1999a: 4), implying that an important appeal of the environmental capital methodology (and by extension, capacity studies) lies in the promise of technical rationality. But the assessments were also acknowledged by those participating in them to have been 'a valuable learning experience' (*ibid.*: 4).

Even if planning authorities have disclaimed any intention 'to give a precise amount of growth that could be accommodated' (Counsell 1999b: 55), they have typically drawn upon capacity studies to argue that development above a certain level would be unsustainable.[6] Critics (noting the preference for this technique in relatively affluent parts of England) have been swift to accuse planners and environmental groups of appropriating scientific authority in defence of politically motivated policies. But the solution, for such critics, lies in recourse to even greater technical rationality. Grigson (1995: 12, 14), for example, argues that capacity studies should involve 'a systematic and comprehensive appraisal', using a common methodology and based on standards with a 'scientific basis', preferably established by a body external to the planning authority. This is curious, for while ostensibly neutral techniques can mask arbitrary decisions (and this may be part of their appeal), 'objective scientific analysis' is not the only, or even the most helpful, alternative. What matters is that policies be defensible, which may require deliberation and judgement as well as science. In any case, it is difficult to think of any form of 'scientific analysis' in planning (including the social surveys that Grigson advocates) that is not itself prone to hidden assumptions and capture. As Wynne (1975: 111) reminds us, 'even innocent data collection is tacitly *guided* by a previously assimilated framework of assumptions and explanatory commitments'. The way forward would seem to lie not in stepping up the analytical arms race but in stimulating a more explicitly political debate.

For the fact remains that delineating environmental capital and making judgements about capacity are not just technical exercises but (however else they may be used) also offer ways of mobilising stronger interpretations of sustainable development, giving primacy to (at least some) environmental considerations. If, in this sense, they emphasise 'the conservationist and preservationist tendencies of the planning system' (Rydin 1998b: 756), they also have the potential to do more: they are, in effect, located on the front line in the battle for different conceptions of sustainable development. At times, this has seemed an unequal contest, not least because environmental claims have to be robustly defended in the planning process but preferences for many goods and services, 'although actually based on judgements open to critical evaluation, [are] treated or responded to as if they were not' (Keat 1997: 37). While Grigson (1995: 25) not unreasonably urges transparency in the use of environmental capacity concepts – 'black box studies that cannot be understood are not evidence' – many of the sectoral policies challenged by such concepts are themselves cloaked in dense technical rationality (Cowell and Murdoch 1999) or based on powerful presumptions about property rights and what constitutes 'the national interest' (Jewell 1995; Owens 1997). The 'black boxes' of demand projections and markets have not traditionally been held to account, with the effect that 'calculations of future demand tend to arrive in local contexts in a form that makes them almost irresistible' (Murdoch 2000: 509). Thus, for example, West Sussex County Council faced legal challenge when policies in its structure plan, based on an elaborate analysis of the county's environmental capacity (Connell 1999; Counsell 1999b, 1999c), provided for lower levels of housing development than those set out in regional planning guidance (DoE 1994c). Although these provisions had survived the examination in public, the High Court upheld the Secretary of State's right to direct what the housing apportionment should be. Rationality seemed to yield to power (Flyvbjerg 1998), or at least to the authority of central government and the alternative rationalities embodied in household projections.[7]

Nevertheless, capacity studies have exerted a kind of discursive power as metaphor, helping to 'shape political argument over society's relationship with the environment' and capturing the idea that 'society must learn to live within limits' (Jacobs 1997b: 67). In highlighting tensions between environmental protection and the development trends implicit in strategic policies, they undermine conceptions of sustainability that depend on 'balancing' or 'integrating' competing priorities.[8] It was inevitable, therefore, that capacity concepts would lead to bruising encounters with development interests (played out in planning inquiries in the UK throughout the 1990s), and that critics would emphasise the potential for 'misuse' (DETR 1997a: 60) and for undesirable distributive impacts; even some of its proponents have suggested that environmental capacity fails 'because it is too strong and inflexible' (Levett 1998:7). The charge of 'environmental absolutism' (Grigson 1995: 17) did lead to a tactical retreat, with many practitioners tempering their language at least. In the aftermath of the High Court challenge, for example, West Sussex outlined 'Strategic Development Options', seeking 'the least un-sustainable locations for necessary development' and informing debate about 'the limits of acceptable change' (Connell 1999: 9). Environmental capital assessments have also become more nuanced, as we have seen, and have been applied relatively safely to topics such as land characterisation and appraisal of alternative sites for development. Even so, the deployment of capacity concepts can be said to have prised open some 'black boxes', shifting the language of housing policy (following conflict over regional planning guidance for the south-east) from 'predict and provide' towards 'plan, monitor and manage' (DETR 2000b; Murdoch 2000), and challenging the primacy of demand projections in minerals planning, an issue that we consider in more detail in Chapter 7.

However, we cannot leave the issue of capacity without touching upon a question that is invariably raised in this context. Surely, some would argue, there is already a basis – in the various methodologies for monetary valuation and in cost–benefit analysis – for dealing with vexed questions of value in assessments of environmental capital and capacity, and with relative priorities when there are conflicts with economic (or social) aspirations. For some analysts, as we have seen in Chapter 3, the use of techniques to quantify preferences for environments and include them in aggregative welfare calculations might be constitutive of the *definition* of critical environmental capital, or of what should be held 'constant', and thus of particular (utilitarian) conceptions of sustainable development. Such an approach is certainly invoked by the notions of 'balance' and trade-off, which remain firmly entrenched in policy-making communities. We now turn, therefore, to look at the forms of technical rationality most closely associated with these ideas.

A common language? Costs, benefits and environmental capital

The formal techniques of valuation and cost–benefit analysis have had less impact in land-use planning than in some related areas of policy,[9] although monetary valuation has sometimes been used in setting environmental taxes that can be claimed to support planning measures.[10] We do not attempt here to review an extensive and vigorous academic debate about the merits, shortcomings and methodological refinements of these techniques. Two points are worth noting, however. One is that the most robust critiques have focused on the ethical underpinnings of cost–benefit analysis and on the non-dialogical quality of an approach in which different moral claims are reduced to preferences measured along a single (monetary) metric (see, for example, Clark *et al.* 2000; Holland 1995; Kelman 1990; J. O'Neill 1993; Sagoff 1988). The other is that persistent unease has had some effect in shifting emphasis towards broader forms of appraisal, such as multi-criteria analysis (Dodgson *et al.* 2000; Martinez-Alier *et al.* 1998; Stirling and Mayer 1999) or the use of valuation techniques in a more dialogical setting (Turner *et al.* 2000). That such refinements cannot resolve all of the issues is illustrated by the new approach to appraisal of transport schemes, to which we return in Chapter 5. Still, the language of balancing, or weighing, conflicting considerations retains a powerful appeal in planning (Lichfield 1994; Moroni 1994). Protagonists in development conflicts have often felt the need to show that environmental protection would outweigh the benefits of development, and their frequent failure in this respect has contributed to a feeling that the only way for the environment to win the battle of political visibility is to reinforce environmental values in quantifiable, monetary terms. Occasional decisions might support this argument,[11] but (as we argued in Chapter 3) environmentally favourable outcomes cannot be guaranteed unless *a priori* constraints have been introduced (Lichfield 1994; Pearce 1993): this was an important reason for the support by conservation agencies and environmental groups of capital- and capacity-based approaches (and, tacitly, of non-utilitarian ethical frameworks).[12]

Although not often applied in statutory land-use planning, monetary valuation and subsequent weighing up of costs and benefits have been used in ways that amount to deciding whether or not particular environments are 'critical'. Such applications have been controversial, not least because notionally methodological choices mask important political and ethical assumptions. At a micro level, the choice between asking for people's 'willingness to pay' to retain or acquire some valued environment, or their 'willingness to accept' compensation for its loss or denial, makes major presumptions about rights and entitlements and dramatically affects the responses given (Hanley and Spash 1993). The choice of constituency is also crucial. In a widely cited case in 1996, which for many epitomised the fragility of valuation techniques, the Environment Agency sought to protect the wildlife and conservation value of the River Kennet in Wiltshire by reducing licensed levels of water abstraction from a nearby borehole (ENDS 1998a). At an estimated net present value of £13.2 million, the 'existence value' of the Kennet was claimed by the Agency to outweigh the additional costs of obtaining supplies elsewhere. But the inspector at the public inquiry accepted the water company's much lower non-use value of £300,000, tipping the economic balance in favour of maintaining abstraction. Leaving aside for now the framing of the issue in terms of meeting rather than managing demand, the difference in the estimates was due to assumptions about the size of the constituency over which willingness to pay should be aggregated. The Agency included all three million households in the water company's supply area, but the company, and the inspector, thought it appropriate to include only people living in the more immediate vicinity, effectively denying that the Kennet could be valued as part of a river system by a wider community (and thus making something of a nonsense of the concept of existence value).

The experience of trying to estimate the existence value of the Kennet bodes ill for the use of valuation techniques to determine whether particular assets constitute 'critical environmental capital'. Refinement of methodologies will not avoid the problem that such techniques beg (or misrepresent) the fundamental questions that deciding what is critical must entail. Since they mobilise predominantly instrumental and utilitarian conceptions of sustainability, which are problematic and contentious for reasons that we elaborated in Chapter 3, they are unlikely to provide sound ammunition in planning conflicts; nor are they likely to prove reliable for those defending an environmental case.[13] Rather, the use of valuation and cost–benefit analysis on the grounds that they will lead to 'the right' substantive results tells us much about the relationship between rationality and power, and certainly belies the idea that such techniques 'merely inform' the decision-making process.

Winning the race?

In fact, whichever means of assessment is selected, its conclusions rarely provide a technocratic trump card in the political arena – either because the basis of the analysis itself is open to dispute or because its proponents lack other forms of political power. Just as valuation techniques conspicuously fail to command authority, environmental assessments or capacity studies are likely to be challenged, however rigorously conducted. It is also clear that none of the techniques discussed above is simply a means of informing decisions to make them 'more sustainable'; some evidently have the potential to mobilise distinctive conceptions of sustainable development, and all are open to capture. Nevertheless, methods allowing scope for deliberation about social priorities would seem to provide a more appropriate framework for 'planning for sustainability' than technocratic, non-dialogical approaches. This would suggest a move away from cost–benefit analysis and mechanistic forms of project-based assessment towards some combination of strategic environmental appraisal and capacity studies. Even if these latter techniques have themselves been driven by the 'analytical arms race', they have shown at least some potential to open up fundamental issues and may yet elude concerted efforts to contain or deflect such scrutiny.

Engaging 'the public'?

Our argument so far has suggested that at least some procedures with their roots in technical rationality may provide important opportunities for dialogue about the nature of sustainable development. But many commentators have called for more widespread use of processes that are both deliberative and participatory in their basic intent, advocating development of consensus conferences, citizens' juries and panels, focus groups and neighbourhood forums, together with techniques that might stimulate wide-ranging debate (for a useful review, see Audit Commission 1999). Reflecting dissatisfaction with traditional approaches, such thinking has reinvigorated the long-running experiment with participation in planning and has stimulated action in related spheres such as Local Agenda 21.[14] In some cases, such initiatives have gone hand in hand with the use of more formal techniques – 'visioning' in conjunction with indicators, for example (Brown 1998; Brugman 1997; Dean 1995), or stakeholder forums with environmental audits (Davoudi *et al.* 1997). The underlying premise of all of this activity is that more inclusive public involvement is 'essential for a truly sustainable community' (DETR 1999c: para 7.87), although rigorous defence of this proposition is less commonplace than its repeated assertion.

One persuasive argument, as we have shown, is that defining what is 'sustainable' must be a deliberative, not simply a technical exercise, and for many commentators it is a natural extension of this view that deliberation should be inclusive. It is seen as a matter of democracy itself, and of justice, that people be involved in decisions that will affect their lives (see, for example, Petts 1999).[15] Bound up with this agenda, and reflecting a wider questioning of scientific rationality and the authority of experts, is a conviction that public values and local knowledge could enrich (or challenge) conventionally recognised expertise, producing results that go 'beyond the capabilities of authoritarian or technocratic methods of policy-making' (Majone 1989: 2; see also Grove-White 1997; Healey 1997; Irwin and Wynne 1996; Macnaghten et al. 1995; Owens 2000; RCEP 1998). These democratic and epistemic rationales have become intertwined with more instrumental ones, some of which perpetuate elements of legitimation: participation must be 'seen to be done', or the public must be engaged in order to be converted to a more 'objective' view, or made more pliable (ESRC 1999; Owens 2000). Public consent is seen as a prerequisite for the macro-level policy changes needed to promote sustainability (Blowers 1993b), while consensual local solutions are deemed to require the ownership of affected parties (DETR 1999d; Parker 1995; Selman 1996). Thus, for example, involvement of a wide range of interests in capacity studies has been urged as a means of making their outputs 'both valid and acceptable' (DETR 1997a: 56).

Whatever the rationale, it has been widely assumed that new forums are needed if the acknowledged difficulties of meaningful participation are to be overcome. It is worth reflecting, therefore, on the adequacy, or otherwise, of established statutory arrangements for public involvement. Broadly speaking, attempts to engage the public – or at least a wider public – in plan making, through the usual methods of consultation drafts, exhibitions, public meetings and inquiries, have enjoyed only limited success (TCPA 1999). Among the most prominent grounds for this verdict have been the persistent failure to reach certain groups (often those who are socially disadvantaged) and the tendency for exercises to be ritualistic, offering 'the shadow rather than the substance' (Hall 1992: 246) of participation.[16] Even so, it would be misleading to imply that traditional arrangements have been wholly without effect. On the contrary, the apertures for 'public' involvement prised open over the years have admitted previously excluded interests, particularly environmental coalitions, edging the policy process in a more pluralist direction. Public inquiries, for example, have brought together diverse groups and individuals, raised the profile of important issues and provided a forum for evidence, argument and critical challenge. Although the reality falls a long way short of deliberative ideals,[17] individual developments have been subjected to strong and effective challenge at inquiry, and successive challenges have contributed, in some instances, to a process of policy learning and change. In effect, procedures intended to give local people some right to be heard have been used in lieu of a wider deliberative process in the polity as a whole; we see this clearly in the context of the roads programme, and of minerals policy, examined in later chapters. Opportunities for participation have been promoted and defended by critics of government policy essentially for this reason, while development interests have persistently called for the 'streamlining' of the planning process, and for the remit of inquiries to be confined to local issues of siting, design and impact management (Owens 1985; see also DETR 1999e). If the *status quo* often proves resilient to challenge, and public involvement seems ritualistic, this might be attributable not to procedural shortcomings but to wider structural commitments to particular patterns of growth.

Whether new approaches to engaging the public will prove more or less successful than traditional techniques depends on the criteria for measuring success, and (in turn) on the rationale for wider involvement. Some deliberative and inclusive forums have undoubtedly been of value for participants and have arguably been more constructive than conventional ways of eliciting public views (for examples, see RCEP 1998; see also Selman 2001). Local Agenda 21 initiatives are interesting in this context, given the enthusiasm for promoting sustainability within an inclusive localised framework and the fact that these exercises have often engaged in some way with the planning process (Davies 1998).[18] Inclusion remains elusive, however. Although some Local Agenda 21 exercises have claimed to engage a broad spectrum of social and economic groups (Cartwright 1997), and there has been interest in novel approaches, one study found that 'few authorities ... employed techniques markedly different from those which have provoked little participation before' (Buckingham-Hatfield 1997: 216).[19] Commentators fret about a general inability to reach beyond the 'green ghetto' (Young 1996: 26) or to move outside policy areas where sustainable development is already an established concern (Littlewood and While 1997); engagement of business interests has been relatively limited, for example, potentially marginalising the strategies emerging from Local Agenda 21 groups (Carter and Darlow 1997; Selman 2001).

Whatever the concerns about who is or is not included, a key issue is that deliberative exercises such as Local Agenda 21 are always likely to involve far fewer people (albeit more intensively) than the established democratic institution of voting, even when (as in the UK) there are dismally low turnouts in local elections (Levett 1999b). Their key role, therefore, is seen to lie in engaging stakeholders and generating ideas to complement and enhance the formal democratic process. Paradoxically, however, it seems that if emerging ideas are to be acceptable, they must be confined to 'safe' issues that do not represent a significant challenge to established interests and norms (Young 1996). When sustainability strategies have transgressed such boundaries, they have been marginalised. In Manchester, for example, a draft Local Agenda 21 statement immediately ran into difficulty because its proposals for civil aviation, including taxation of air travel, clashed with powerful city council support for Manchester Airport with its associated employment (Kitchen 1997). Similarly, on land-use planning issues, ideas expressed through community involvement tend to be rationalised by offi-cers in terms that they regard as institutionally legitimate (Holt-Jensen 1997), and the extent to which Local Agenda 21 has actually influenced development plans is unclear. Lack of direct influence must sometimes be attributed to the proper role of a democratic process in which other legiti-mate considerations prevail. But it might also be explained (as might the limited participation of business interests, noted above) by the remoteness of these initiatives from the real levers of economic and political power.

While it would be premature to reach definitive conclusions about delib-erative and inclusive processes, it is appropriate to identify questions that merit more rigorous attention, particularly in the context of sustainable development. One of the most difficult – that of discursive competence – tends to be evaded, perhaps because it invokes accusations of elitism, yet it is a question that must be confronted, particularly where complex and demanding environmental and social issues are involved (Foster 1997). It has been difficult to disentangle calls for deliberation (as an alternative to arriving at outcomes through technocratic or non-dialogical processes) from advocacy of inclusive or 'stakeholder' participation, counterposed with deci-sion making by 'experts'. Dialogue, or deliberation, can take place at any level of decision making and may involve few or many people (expert or otherwise); 'participation', on the other hand, is not necessarily deliberative, and many forums have been far removed from Habermasian 'ideal speech' situations. It is ironic – or perhaps it is symptomatic – that the urge to cast the net more widely comes at a time when the demise of civic virtues has been almost universally acknowledged and lamented. Yet many advocates of deliberative and inclusive processes seem to follow Sagoff (1988) in seeing it as relatively unproblematic for participants to cast off the values of market consumerism (at the door, as it were) and to be instantly ready to 'deliberate' as citizens, in possession of all the necessary information, skills and disposi-tions. The risk that inclusive deliberation will produce little more than a chaos of voices can be lessened by dedicated input of time and information, but there is then a difficult line to be drawn between providing 'neutral' support and steering the debate, and the practicality of generalising such resource-intensive processes becomes an issue.

If deliberative forums could, with patience and good practice, be brought closer to some ideal (and arguably might help to inculcate civic virtues[20]), questions about the wider legitimacy of outcomes reached in specific (often locationally specific) contexts remain. Even exemplary processes may lead to particularist interpretations of sustainable development (Dobson 1998; Jacobs 1997a), undermining the very concept that is normally held to transcend localities and incorporate 'distant others' (Agyeman 2000). The dilemma is acutely present in planning for sustainability, often strongly associated with 'the local' or 'the local community' (Littlewood and While 1997) but inevitably entailing conflicts of interest and obligation between different places and different scales. However inclusive the process through which sustainability is defined, local communities cannot be permitted a monopoly of interpretation simply by virtue of being local. On the other hand, there are dangers in too rigid a hierarchy, assuming automatic precedence for higher scales of political authority. Not only does local knowledge provide a valuable and sometimes vital perspective, but opportunities for 'local' involvement have become important apertures through which prevailing conceptions of sustainable development (and the 'wider public interest') can themselves be subjected to scrutiny and debate. The sharpness of these issues is exemplified in later chapters, perhaps most clearly in our discussion of minerals planning, but the point to be emphasised here is that they cannot be resolved simply by repeating the mantra of 'involving the local community'.

Even if the key issues could be confined to a single scale, one might question how far deliberation can reasonably be expected to converge upon agreed outcomes. Advocates of new procedures have placed a strong emphasis on consensus, apparently believing that dissent associated with sustainable development is 'unproductive' (Selman 1996: 78). But to argue that important issues should be approached through dialogue is not to say that all issues can thereby be rendered consensual. Indeed, an undue emphasis on consensus may 'smother ... the passion of ideological conflict' (Jacobs 1997c: 4) and, if the starting point involves significant inequalities of power, might lead to outcomes that are unjust, even when all parties consider themselves better off. None of this is to deny that consensus-building approaches have potential, in certain situations, to forge agreements that represent more than narrow political compromises (Floyer-Acland 1990; J. O'Neill 1993; Susskind 1981; Susskind et al. 1999). Learning can take place, interests are not always fixed and immutable, and areas of dispute can sometimes be narrowed or eliminated. However, certain principles may not yield easily to such treatment, and conflicts over sustainable land use often violate the contextual conditions for consensual solutions: that parties agree that the current situation is unacceptable; that they accept the 'need' for some facility; or that conceptions of fundamental rights are not involved (Susskind 2000). There is a danger, in such circumstances, that the search for consensus will become a means of suppressing weaker voices and defusing legitimate political activity. In the new enthusiasm for deliberation, we should not lose sight of the possibility that 'forms of participation that are practical, committed and ready for conflict provide a superior paradigm of democratic virtue ... [to those] ... that are discursive, detached and consensus dependent' (Flyvbjerg 1998: 236).

Finally, in order to address widely neglected questions of agency, it is important to locate deliberative and inclusive approaches to planning within their political and institutional context. It is one challenge, as we have shown, to demonstrate that new ways of engaging the public can produce insights and ideas that make a legitimate contribution to some wider democratic process, but positive responses to this challenge simply beg another question, one that is familiar to students of traditional approaches: if outcomes favour radical policy change, how successfully can they confront the unsustainable practices perpetuated by dominant forms of economic power? Initial experience, at least, suggests that new approaches can rarely undermine the capacity of economic interests and bureaucracies to capture the agenda and frame what is to be regarded as politically acceptable: as Campbell and Marshall (2000: 340) observe, simply giving voice to different forms of knowledge need not mean that 'the pervasive structures in society will be challenged'. There is even some evidence that initiatives might be counterproductive if their role and relation to the formal democratic process are unclear. In a study of transport planning in Munich, for example, Hajer and Kesselring (1999) found that the introduction of 'add-on' interactive forums had 'by no means' facilitated a move towards greater sustainability in the short term (*ibid.*: 17) but 'might have eroded the power of "official" [legally embedded] practices of participation' (*ibid.*: 19).

What emerges from this discussion is that relationships between participation, deliberation and sustainable development need more careful thought, and any analysis must be mindful of the structures of economic and political power within which these relationships are established. We have suggested that traditional arrangements for public involvement in planning, although much criticised, have acted as crucial conduits for the introduction of new ideas and different world views, including, latterly, challenges to official conceptions of sustainable development. Deliberation has indeed taken place, fostered not by tailor-made procedures but by open political conflict over development proposals and the policies that lie behind them. Policy learning may be evident too, although (as we shall show in ensuing chapters) challenging dominant interests is invariably a slow and painful process. New approaches to 'engaging the public' seek to be more inclusive and accessible than the old, and (in many cases) to produce consensual strategies for sustainable development. Given time, deliberative and inclusive forums might themselves introduce new perspectives and ideas into the planning process; experience suggests that this is not something that can be judged in the short term. However, initial evidence does not neatly support the view that these approaches yield more legitimate or more influential results than traditional ones, and the emphasis on inclusion and consensus may have deflected attention from other important issues. If the failures of meaningful participation arise not from the confrontational or non-inclusive nature of particular procedures but from the wider political and cultural context of which those procedures are a part, it may be an illusion to think that anything better could be achieved by procedural innovation alone. This calls for at least some caution, since an uncritical rush towards new forums risks undermining the hard-won legitimacy of existing statutory arrangements.

We conclude this section with a few reflections on the claim that participation is essential for a truly sustainable community. Deliberation and good judgement are, we have argued, integral to the interpretation of sustainable development, but deliberation and inclusive participation do not automatically go together; nor do deliberative and inclusive processes necessarily produce outcomes compatible with broader principles of environmental or social justice. The conceptions of sustainability mobilised by particular processes will depend on the context, on how the issues and possibilities are framed and on who gets involved – a contemporary example of Schattschneider's (1960: 71) 'mobilization of bias'. Whether these conceptions are influential will depend on their legitimacy in a wider political context, and on the extent to which they are consonant with powerful economic and political interests. The significance of many of these issues will be emphasised in later chapters. Now, however, we turn to the distinct but related agenda of 'integration', another set of reforms widely regarded as conducive – even essential – to the promotion of sustainable development.

Integration: the elusive goal

As Downs *et al.* (1991) observe, integration is a term with multiple meanings. In the context of sustainability, it usually implies co-operation or co-ordination between different organisations, sectors and levels of government, and it sometimes requires significant institutional change. Horizontal integration, between different parts of the state apparatus (and, increasingly, the private sector), has been central to the discourse of sustainability, typically with a view to ensuring that environmental objectives are incorporated into decision making in vital sectors such as transport, energy and agriculture (see, for example, Farmer *et al.* 1999). Arguments for integration on a territorial basis, evident in the European Spatial Development Perspective and various regional planning initiatives, have also gained ground. Institutional integration, at the very least involving co-operation between different interests and organisations, has usually been regarded as indispensable in promoting all of the other kinds.

While the rhetoric of integration is pervasive, its precise meaning and its implications for policy practice are often unclear. So are the mechanisms through which it might promote sustainable development, although the appraisal and audit techniques discussed earlier in this chapter, as well as participatory procedures, are usually assumed to be conducive to integration,

and therefore to greater sustainability. Given the uncertainties, it is important to examine the 'policy work' (Healey 1998b: 3) that calls for greater
integration perform. Who is being asked to integrate what, with whom and
how, and what conceptions of sustainable development are different parties
being invited to share? These questions, and the success or otherwise of integration in its various senses, are particularly interesting when we deal with
specific policy sectors. Our discussion in Chapter 5, in particular, shows how
'integrated transport' provided a convenient (but ultimately unstable) storyline around which different interests could rally. Here, however, our aim is
to make more general observations about integration in a territorial or
spatial context, noting how often that land use or spatial planning is deemed
an appropriate focal point for the co-ordination of sectoral policies (Farmer *et
al.* 1999; Janssens and van Tatenhove 2000) and charged with the task of
reconciling economic, environmental and social programmes.

Integration in a spatial framework

The Dutch ROM initiative

That differences between rhetoric and reality may be substantial, even in
states regarded as leaders in sustainable development, is illustrated by Dutch
experience with a holistic approach to planning during the 1990s. The so-
called ROM initiative sought to integrate land-use planning (*ruimtelijke
ordening*) and environmental planning (*milieubelied*) into a regional framework
and was initially applied in eleven areas.[21] The projects were always
intended to be participatory and deliberative: national, provincial and local
tiers of government were drawn together with public and private interest
groups around production of a vision for each area. This was to be accompanied by an action plan detailing specific objectives and projects, and an
administrative agreement allocating responsibilities between the various
parties (Bouwer 1994). The hope was that involving a broad range of participants, and including economic and social as well as environmental
considerations, would forge agreement on innovative solutions to entrenched
environmental problems (Glasbergen and Driessen 1994). 'Integration', in
this model, would help to deliver sustainable development at the regional
scale (Netherlands Ministry of Housing, Spatial Planning and the
Environment 1998).

After the initial experiments, participants in the ROM initiative
(including the Dutch government) remained positive about its prospects. A
number of projects had been implemented successfully, with those aimed at
economic development faring best but modest environmental enhancement
also being achieved in some regions (Driessen and Glasbergen 1995;
Janssens and van Tatenhove 2000). Commentators also noted improved
communication between different actors, better prospects for achieving

consensus and elements of learning (Bouwer 1994; Glasbergen and Driessen 1994; Netherlands Ministry of Housing, Spatial Planning and the Environment 1998), all of which can establish fertile ground for reframing issues and policies in the longer term. With a little more experience and hindsight, however, it seems that some of the early optimism might have been misplaced: progress with regional integration has been slower than expected and accountability poor; indeed, weakly developed and largely aspirational goals have made the monitoring of achievements difficult (P. Glasbergen, personal communication 1999; see also Janssens and von Tatenhove 2000; Netherlands Ministry of Housing, Spatial Planning and the Environment 1998). Instead of smoothly integrating different priorities and thereby promoting sustainable development, the ROM projects have often exposed the kinds of tension that we identified in Chapter 3, notably that between balancing 'the interests of various societal activities' and securing national standards or targets for environmental quality (Bouwer 1994: 109; see also Netherlands Ministry of Housing, Spatial Planning and the Environment 1991a; van der Gun and de Roo 1994). In the Netherlands (as in the UK) these different approaches have become institutionalised in systems of land use and environmental planning respectively, creating a barrier to integration within any spatially defined framework. In some instances, environmental standards that seemed at odds with social or economic goals stimulated 'coherent resistance in local communities' (Bouwer 1993: 4, 1994), so that environmental protection proved difficult to reconcile with commitments to consensus and partnership. Elsewhere, environmental objectives had to bend to economic agendas that could not be renegotiated by ROM partnerships (Bouwer 1994). In the case of the Schiphol area, for example, a consensual vision had to accommodate the 'national interest' in growth of the airport, with concomitant threats to local air quality, noise pollution targets and national emissions objectives (Glasbergen and Driessen 1994: 36). Here the outcome of 'integration' appears in practice to have been the least-cost accommodation of growth.

Generally, the ROM experience suggests that even concerted efforts at sectoral integration may not easily align different conceptions of sustainable development; if anything, they have mobilised weaker versions of sustainability and exposed the shallowness of Panglossian interpretations. Bringing different aspects of planning together in a stakeholder forum cannot of itself (and certainly not in the short term) exert leverage over powerful external actors, reconcile divergent beliefs about problems and solutions, or overcome structural contradictions between economic and environmental goals (Jessop 1997a). If such initiatives produce lasting benefits, these are most likely to be diffuse ones, emerging gradually from the process of learning within and between coalitions in the medium or longer term. Even then, changes external to the planning system and to the region will be needed if development is to become sustainable in a radical, or stronger, sense. The more immediate verdict on integrated regional planning is that 'good intentions, focusing on processes of negotiation and cooperation, prevail over substantial and sustainable development' (Janssens and van Tatenhove 2000: 169).

Integrating objectives in UK development plans

Although there has been no formal planning initiative in the UK to match that of the ROMs, the experience of trying to integrate environmental, social and economic considerations in development plans does have some parallels with that of the Dutch experiment. Land-use planning, as we have shown, has increasingly been expected to become an instrument of sustainable development, and regional planning guidance has moved closer towards becoming a spatial plan in which sectoral policies with land-use implications can be co-ordinated. The urge to 'integrate' – to achieve multiple objectives – is effective up to a point. Coverage of environmental and social issues in development plans has improved, in part because of requirements for environmental appraisal, and policies with adverse impacts have sometimes been rejected as a result (see, for example, Davoudi *et al.* 1997). But it is not always possible to meet diverse objectives simultaneously, and when conflict is unavoidable it becomes difficult to see what 'integration' means in practice beyond the inclusion in plans of mutually incompatible goals. Economic development still tends to be seen as the primary objective in regional planning, despite rhetorical support for the principles of sustainable development (University of Hull Institute of City and Regional Studies 2000), while local authorities have been unable or unwilling to eliminate growth-oriented planning policies even when they conflict with a range of environmental objectives. Road schemes, as we show in Chapter 5, are a particular case in point, often enjoying substantial local and regional support with sufficient political authority to override critical environmental assessments.

As in the Netherlands, while integration is promoted rhetorically as an *instrument* of sustainable development, an important barrier to its achievement in planning practice is that environmental and economic networks promote different, and often incompatible, conceptions of what sustainable development entails (Clark *et al.* 1993; Davoudi *et al.* 1997; Gibbs *et al.* 1998; Healey and Shaw 1994; Myerson and Rydin 1994). Economic development coalitions are unlikely to accept sustainability as a challenge to the very nature of economic activity;[22] for them, integration means (if anything) taking better account of environmental considerations and enhancing environments in order to encourage investment – environmental capital in the service of economic capital. Environmental coalitions favour stronger conceptions of sustainable development, in which environments are regarded as 'foundational' (Jacobs 1997b: 31): integration in this framework means modifying economic behaviour to take account of environmental constraints.

Divergent perspectives were apparent, for example, in the protracted conflict over the redevelopment of Cardiff Bay in South Wales, an issue that we consider in Chapter 6. Local economic interests stressed the city's vulnerability in a global economy, and the need to mould environments and social conditions to create stable market opportunities; conservation groups emphasised the fragility of the estuarine ecosystem and preferred to frame economic priorities in relation to environmental risks (Cowell 2000a; Thomas and Imrie 1999). Such positions are familiar to students of land-use conflicts; what is noteworthy is the apparent failure of those calling for 'integration' to grasp how deeply rooted are the differences in beliefs and interests involved.

Perhaps the best hope (as with the ROM projects) is that channels of communication will be opened by attempts to pursue different objectives within a more integrated planning framework, and that this might ultimately lead to changes in prevailing concepts of development. To date, however, the evidence for fruitful dialogue of this kind is rather slim. Environmental interests have been inclined to pursue integration *within* the environmental sphere, forming 'sustainability' networks centred on plan making, while separate partnerships promote a parallel economic development agenda (Davoudi *et al.* 1997; Selman and Wragg 1999). We suggested earlier that requirements for policy appraisal have the potential to bring different interests together and stimulate debate about divergent goals (see also Gibbs *et al.* 1998), and in this sense appraisal might be seen as an important means of initiating communication and co-ordination. However, environmental appraisal of development plans has not − or has not yet − lived up to such expectations. It has often been ritualistic, relatively closed and (especially in its expanded, 'sustainability' format) seems more likely to reinforce the pre-eminence of traditional growth discourses than to locate the environment as a basic context for economic development. Nor is the relationship between more integrated, 'joined-up' policy making and greater public participation straightforward. Where effective integration and joint working require high-level brokering, the need for sensitive negotiation can militate strongly against public involvement (Wood 1989; see also Reade 1982).

The real issues

It is difficult to avoid the conclusion that enthusiasm for integration has been well intentioned but naive. As with sustainable development more generally, the impetus has come largely from environmental interests concerned about the absence of environmental considerations in key policy areas. This seems to confirm Degeling's (1995: 295) observation that 'calls for better policy coordination across or between sectors generally emerge as part of (and in the context of) sectoral politics ... as actors within one sector attempt to get actors in other sectors to take on aspects of their concerns'.

Usually, however, when environmentalists call for integration, they see it as a way of mobilising stronger conceptions of sustainability, believing that if environmental considerations are made integral to (say) economic development then the latter must be transformed. On the other hand, economic interests, and governments, have generally operated with different assumptions: they have tended to accept only those models of integration that mobilise weaker conceptions of sustainability (albeit taking 'better account' of the environment), or Panglossian versions in which (with a little institutional re-jigging) previously conflicting objectives somehow become mutually interdependent. Thus the UK government's tripartite definition of sustainable development can be used in regional planning 'to promote economic development and competitiveness as ends in themselves' (University of Hull Institute for City and Regional Studies 2000: 1), in effect reducing the new discourse to little more than a rationale for business as usual. This also explains why 'sustainability appraisal', rather than the environmental kind, has become the officially favoured tool of integration.

Degeling's (1995) critique of simplistic models of 'intersectoralism', although written with regard to a different empirical context,[23] contains much of relevance to our discussion of integration as an instrument of sustainable development. Appeals for intersectoral co-ordination, Degeling suggests, ignore the ways in which sectors themselves are constructed and maintained by particular forms of knowledge and expertise, well-defined policy territories and patterns of resource allocation. Such 'modes of sectoring' are too deeply ingrained to be overcome by 'good intentions, snappy commonsense thinking or some optimum design fix' (*ibid.*: 295). In a similar way, calls for 'integration' in planning fail to recognise deep differences between the sectors or objectives that are to be treated in this way. Not surprisingly, then, more co-ordinated or integrated approaches 'rarely get beyond exhortation, and those that make a promising start often end in the sand' (*ibid.*: 290). Our planning examples seem to confirm that even when integrative mechanisms are in place, 'the structuring processes that are integral to existing modes of sectoring remain virtually untouched' (*ibid.*: 295): economic and environmental coalitions do not suddenly begin to share compatible conceptions of sustainable development in their area, for example, or relinquish their normal constituencies or lines of influence. Like Degeling, we do not deny that existing, fragmented approaches may be problematic – after all, development has been far from sustainable under almost any definition – but we suggest that too much has been expected of administrative or technical approaches to integration when the real issues are those of power and advantage. The way forward, according to Degeling (*ibid.*: 300), will be found 'not in the search for some bias-free universalism, but in creating room for manoeuvre for new biases and agendas'. The intriguing question is whether the various integrative initiatives in planning, although clearly subject to the structural constraints that Degeling describes in the short term, might nevertheless provide a forum within which radical new agendas can develop.

Concluding comments

Drawing together our thoughts on the technical, participatory and integrative approaches considered in this chapter, it is clear that whatever their individual merits or shortcomings, none can be said to lead straightforwardly to 'more sustainable development'. Indeed, an important conclusion is that to conceive of these approaches as *instruments* in promoting some preformed, consensual concept of sustainability is profoundly misleading. In practice, all are bound into power struggles in which conceptions of what is sustainable are actively constructed and negotiated. And since these struggles are invariably unequal, no assessment of the role of appraisal, or participation or integration, is likely to be adequate if it is divorced from fundamental questions of agency and leverage over the political process. Our argument is not that the techniques examined have always been without effect (indeed, the degree of resistance to them has sometimes been testimony to their potential), but that their role in informing policies and decisions has been inadequately conceptualised. As a result, more has generally been expected of various procedures than they are able to deliver, while the necessary ingredients of meaningful change remain unhelpfully obscure. Further, although methodological and procedural refinements may in some cases be beneficial, faith in their ability to make a significant difference to outcomes may be deeply misplaced.

The relationship between techniques (broadly defined) and sustainable development is complex and often context-dependent. Some approaches, as we have seen, mobilise distinctive conceptions of sustainability. Capacity studies, for example, are strongly associated with environment-led interpretations, and cost–benefit analyses with utilitarian ones. In other cases, conceptions of sustainable development *emerge from* techniques or processes rather than being predetermined by their underlying premises; thus in many attempts to achieve integration, or inclusive deliberation, what emerges as 'sustainable' depends on the arrangement of actors, opportunities and constraints in any given setting. Sometimes techniques promote only those conceptions of sustainability that are implicit in prevailing policies and institutional arrangements – environmental assessment of projects, which tests for compliance with policy but does not change it, is perhaps the best example of this phenomenon. When, as is increasingly the case, different approaches are employed in mutually supportive ways, much will depend on the permitted combination in a particular political context: if wide-ranging SEA can be used in conjunction with concepts of environmental capacity, for example, it may point towards radical modifications of conventional development trajectories.

A second significant conclusion is that it may be less helpful than is often implied to counterpose 'technical' with 'deliberative' approaches. Our review suggests that most techniques – even the more technocratic ones – provide some forum within which knowledge can be assembled, argument can take place, and learning may occur within and between different coalitions. Although there is much to support the assertion that formal audits and assessments allow political choices to masquerade as technical issues, they do not 'merely' do this. To see only a bogus technical rationality is to fail to acknowledge how (for example) environmental assessment, or capacity studies, can be exploited strategically to challenge deeply embedded assumptions and inject different world views into the decision-making process. Over time, such techniques can thus become part of the 'search for intelligible solutions' (Weale 1992: 222). Even environmental valuation and cost–benefit analysis, conceptually the least dialogical of the approaches reviewed in this chapter, may have some such effects in application – certainly, the River Kennet case led to much reassessment and discussion, even if not immediately to a more palatable outcome. In planning, the ethical and political issues bundled together by sustainable development tend to elude confinement within technical, managerial rationalities, what-ever their intent, and while this may disappoint the 'beleaguered decision-maker' (Petts 1999: 34), it can provide unexpected opportunities for learning and change.

Of the processes that are more explicitly aimed at deliberation and partic-ipation, we have argued that traditional approaches have many shortcomings but have nevertheless provided important apertures for challenge to the *status quo*. This contribution should not be overlooked in the rush to embrace 'deliberative and inclusive processes', whose efficacy and legitimacy still require some clarification. Deliberation and participation may both be necessary in interpreting sustainable development, but their combination in particular circumstances needs more rigorous thought, and it should not be assumed that consensus is invariably more transformative than conflict. For both established and novel arrangements, the most important questions are whether, how and over what timescales they might influence thinking and policies. At root, these are questions – equally applicable to the use of more formal techniques and to initiatives aimed at 'integration' – about the rela-tionship between agency and structure.

Although we have argued that technical, participatory and integrative procedures, sometimes working in combination, offer scope for productive learning and policy change, our evidence also suggests that, in the short term at least, this scope may be severely restricted. A third conclusion – confirming what many others have shown – is that all techniques and proce-dures are subject to capture: they can be manipulated by dominant interests, or deployed in ways that reflect a more subtle exercise of power to ensure that they produce outcomes that reinforce (or do not threaten) established norms. At the same time, it may be difficult for others to challenge the powerful by adopting technical rationality themselves, or by participating in deliberative or integrative experiments. The 'analytical arms race', despite much effort, has produced few clear victories in which valued environments have been protected by better information and assessment alone, and some areas of policy and economy seem able to stay beyond the reach of 'integration' or public challenge. For many issues, the outputs of different techniques and procedures have proved most defensible when they lie safely within both the legislative remit of planning systems and the parameters of the broader political context. 'Planning for sustainability' seems to confirm a longstanding view that structural and contextual factors are a more important influence on outcomes than the technical (or, we might add, procedural) characteristics of the 'instruments' employed (Majone 1976, quoted in Rees 1988: 180).

This leaves us with an interesting and important question. On the one hand, we claim that a range of techniques and procedures, because *inter alia* they provide forums for learning and apertures for new ideas, have the potential to mobilise radical conceptions of sustainable development. Some combination of capacity studies, environmental assessment and deliberation, taking place broadly within a spatially defined framework, looks promising in this respect. On the other hand, it would seem that because outcomes must always be constrained by existing structures of power, the various approaches will be deployed in ways that promote 'safe' conceptions of sustainability, and any that do not will be marginalised. Which of these perspectives dominates must in large part be a question for empirical inquiry. It is one that we address in ensuing chapters, paying particular attention to ways in which the various forms of ammunition are mobilised or disarmed in different planning contexts, and how this process is formative in wider political conflicts over sustainable development. But first we turn, in Chapter 5, to a set of issues that has certainly been the locus for one of the most intense of these conflicts – concerning transport, environment and associated land-use policies. Interestingly, in the context of the foregoing discussion, it was a conflict given form by dissatisfaction with standard assessment techniques and fragmented approaches to policy, and it has been one in which initially limited opportunities for public involvement have been exploited and expanded in a struggle over profoundly different conceptions of sustainability.

5

MOVING TARGETS

Planning for an integrated transport policy

The car has dramatically changed our lives, giving freedom to choose where we live and work, and how we enjoy our leisure.
Automobile Association undated: foreword

The triumph of despotism is to force the slaves to declare themselves free.
Berlin 1969: 165

Introduction

In the last decade of the twentieth century, remarkable shifts took place in thinking about transport policy in many Western countries. At the root of this change lay a conviction that the inexorable traffic growth of the postwar decades, and attempts to accommodate it, were having serious social and environmental impacts. Increasingly, these trends were seen as unsustainable. By the mid-1990s, key international organisations were expressing concern about land-use and transport policies leading to 'excessive travel by car' and agreeing that 'large scale road investment is no longer seen as a solution' (ECMT and OECD 1995: 13). The need to discourage road traffic growth, particularly through 'internalisation of external costs', was a recurrent theme of the European Commission's Common Transport Policy (CEC 1992b: para 39).[1] To varying degrees, individual member states adopted policies with less emphasis on accommodating road traffic and more on reducing the need to travel and promoting environmentally friendly modes; this was as true of countries like the Netherlands, which already had extensive and efficient public transport networks, as of the UK, where the neo-liberal economic policies and permissive land-use planning regime of the 1980s had accelerated the decline of public transport and done much to encourage the 'great car economy'.[2]

In Britain, this 'new realism' (Goodwin *et al.* 1991) had been crystallised, ironically, by traffic projections published in 1989, followed by publication of 'the biggest roads programme since the Romans'[3] (DoT 1989a, 1989b).

The prospect of a doubling or even tripling of the volume of traffic over a thirty-year period, the scale of road building envisaged, and the likelihood of prolonged political confrontation stimulated new thinking and, we shall argue, contributed to a process of policy learning within and between the different coalitions in the transport policy community.[4] The influential Royal Commission on Environmental Pollution entered the fray with a substantial report (RCEP 1994) arguing that transport trends were unsustainable and proposing radical changes to policy, including sharp increases in fuel duty and deep cuts in the roads programme. In the same year, the Standing Advisory Committee on Trunk Road Assessment (SACTRA 1994)[5] confirmed the long-held view of many critics that new roads not only redistribute but also have the potential to generate traffic, sometimes substantially. The UK Round Table on Sustainable Development, set up by the government in 1995, produced several hard-hitting reports on transport in quick succession (UKRT 1996a, 1996b, 1997c, 1997d). Significantly, at a time of mounting concern about climate change, transport was officially acknowledged to be the fastest-growing contributor to greenhouse gas emissions in the UK.

By the mid-1990s, there seemed to be a remarkable convergence of views and widespread adoption of broadly similar discourses about 'integration' and 'choice' in transport policy (Goodwin 1996). The Confederation of British Industry, complaining that successive governments had failed to integrate transport into strategic economic and environmental policies (CBI 1995), sounded remarkably like the Council for the Protection of Rural England, which maintained that transport policies were 'failing the environment … society and the economy' (CPRE 1995: 11). Local authorities, individually and through their associations, had already been calling for integrated and balanced approaches and were emphasising the importance of environmental considerations (see, for example, Association of District Councils 1990; Association of Metropolitan Authorities 1990). Even motoring organisations were forced to confront the adverse impacts of traffic growth. The Automobile Association (AA undated: section 3) acknowledged limits to 'totally unrestrained vehicle ownership' and called for investment in public transport as well as roads (its traditional concern), while the newly established RAC Foundation for Motoring and the Environment (1992: 13) considered it 'inevitable', in the absence of a major technical breakthrough, that greater restrictions would have to be placed on the use of the car.

This reframing of the transport problem was finding its way into the political mainstream, partly through the 'Great Transport Debate' launched by Secretary of State for Transport, Dr Brian Mawhinney, in 1995 (Mawhinney 1995; DoT 1996), and the then Conservative government cautiously accepted the need for some restrictions on road travel, 'if they improve other choices' (DoT 1996: 30). Opposition parties acknowledged the futility of building 'more and more roads to accommodate the projected increase in traffic' (Labour Party 1996: foreword) and promised to develop 'integrated transport systems' to support a sustainable and efficient economy (Liberal Democrats 1995: summary). While a process of learning was arguably taking place in what Kingdon (1995) calls the 'problem' and 'policy' streams, a key event in the 'political stream' – the 1997 general election – provided a window of opportunity to take policy practice forward in line with the new realism.[6] At such moments, Kingdon argues, problem, policy and political streams are merged and lasting policy change becomes possible. Certainly, the incoming Labour government hastened to produce a strategy for integrated transport, published as a White Paper in July 1998. Its opening words proclaimed 'a consensus for radical change in transport policy' (DETR 1998f: 3).[7]

With the important acknowledgement that forecasts would be revised in the light of policies, the Integrated Transport White Paper seemed to consign to history the old 'predict and provide' approach. It promised integration in a number of different senses: between modes, between transport and other government policies (including environmental objectives), and between transport and land-use planning.[8] Local authorities would acquire powers to levy charges on roads and private non-residential parking facilities, and they would be allowed to recycle most of the revenue into public transport and enhanced conditions for pedestrians and cyclists; this would be in addition to a pre-existing commitment to increase the real rate of excise duty on fuel by at least 6 per cent per annum (the so-called 'fuel duty escalator').[9] Emphasis would be shifted from construction of new roads to maintenance of the existing network, and concerted measures would be taken to improve bus and rail systems. Further, since road space and resources were to be reallocated in favour of non car users, the long-established dominance of motorised traffic was to be challenged. Although, as we discuss later, the White Paper failed to confront certain fundamental assumptions underlying transport policy, it would certainly qualify as a 'new realist' document.[10]

The prospects for achieving such radical change and the robustness of the 'consensus' that emerged in the 1990s are important themes for this chapter. However, we do not attempt to cover the whole dynamic and diverse field of transport policy. This has proved fertile ground for students of policy change, including its important cognitive dimensions, and there are many interesting analyses and overviews (see, for example, Adams 1981; Banister 1998; Banister and Button 1992; Dudley and Richardson 1996a; Goodwin et al. 1991; Hamer 1987; Haq 1997; Tyme 1978). Rather, while setting our discussion within this broad framework, we pay particular attention to those aspects of transport policy that relate to land-use planning, which has invariably been seen as a crucial component of new realist approaches. We look first at the connections between transport and land use, some direct and immediate (such as the landscape impacts of new roads), others subtle, longer-term and mediated by social and economic change. Into the latter category would fall the extraordinary upward spiral of mobility and dispersal common to many Western countries in the postwar decades. We then consider the options available to planners for 'integrating' land use and transport as they are urged by almost everyone to do, and we examine critically the policies adopted to date. Here we draw primarily on experience in the UK and the Netherlands, but we also look to some interesting developments in the United States. Finally, we ask to what extent such policies are contributing, or are likely to contribute, to more sustainable patterns of development and movement. We agree with many commentators that land-use planning must be part of a wider policy package if it is to be not only an effective instrument of environmental protection but also to have any lasting effect on travel demand. Even so, we suggest that implementing such a package (as set out, for example, in the UK Integrated Transport White Paper (DETR 1998f)) is likely to remain problematic given the resilience, and dominance, of core values and beliefs about growth, competitiveness and the satisfaction of want-regarding preferences.[11] Thus we return to some of the central themes of this book: the interweaving of knowledge and power in policy change; planning as a forum for learning and its role in the articulation of preferences and values; and the profound challenge of sustainability if it is to mean more than a modest reorganisation of trade-offs between different societal goals.

Transport, land use and sustainability: reframing the connections

The direct impacts of transport systems fall well within traditional planning concerns, although controversy surrounding them intensified with both general environmental awareness and the scale of development taking place. The actual land take associated with transport networks is modest but not insignificant: roads occupy about 3.3 per cent of the land area of the UK, for example, rising to as much as a fifth in urban areas (RCEP 1994). More important have been the often severe effects on wildlife habitats and landscapes, combined with less tangible impacts on cultural and social environments as traffic came to dominate in both urban and rural locations. Mounting protest against expropriation of these different spaces raised the public and political profile of transport issues and became a key factor in the emergence of the new realism. Nor should we forget those impacts on land arising at other points in the materials supply chain: of some 250 million tonnes of aggregates used annually in the UK in the early 1990s, for example, about a third was for construction and maintenance of roads; indeed, these uses have constituted an important part of the 'need' justifying aggregates extraction (to which we return in Chapter 7).

However, of at least equal significance are the subtle and longer-term interactions between transport and land use, a relationship long recognised but one that has attracted renewed attention in the quest for sustainable development. It is widely accepted that transport systems influence forms of development and that patterns of land use affect travel behaviour. But the nature and policy implications of this relationship have been contentious, and the rationale for integrating land use and transport planning has shifted as problems and potential solutions have been framed in different ways. Within the 'predict and provide' paradigm of the 1960s and 1970s, the relationship was represented, albeit in a somewhat crude and deterministic fashion, in large-scale land-use/transportation models made viable by the power of new computers. Subject to much criticism on epistemological and methodological grounds (see, for example, Sayer 1979), such models proved in any case too cumbersome and expensive for routine use, although modelling in some form remains a staple in transport policy and planning.[12]

In 1970, when the new UK Department of the Environment (DoE) brought together the functions of the Ministry of Transport and the Ministry of Housing and Local Government, a key argument for the new 'super-ministry' was to include responsibility for urban planning and transport in one department in the interests of economy and efficiency (McQuail 1994). Emphasis at that time was on providing for travel demands. The Department of Transport was a separate entity again from 1976 to 1997, when the incoming Labour government merged it with the Department of the Environment. In the new realist 1990s, the objective of integration was reframed in terms of sustainability. Indeed, recognition that land use, transport and environment were connected across a range of temporal and spatial scales – that local development control decisions, for example, might ultimately have repercussions for the global climate – was an important factor in establishing the 'new remit' for planning discussed in Chapter 2, helping to legitimise concerns extending beyond the traditional spheres of amenity and the efficient use of land. Within the frame of sustainable development, an understanding of land-use/transport interactions was sought more as a means of slowing or even halting trends than accommodating them, and the whole relationship acquired a new policy significance. Before examining policies, however, we outline the co-evolution of land-use and transport patterns over several decades – a picture broadly discernible in many Western countries – and briefly review the claims that such trends are unsustainable.

Growing prosperity, particularly since the middle of the twentieth century, has been associated with the increasing mobility of both people and goods,[13] permitting geographical dispersal of land uses once more intimately mixed. At the same time, services in both public and private sectors have been 'rationalised' in a quest for economies of scale, leading to fewer, larger facilities often located where they cannot easily be reached by means other than (usually private) motorised transport: hospitals and retail outlets are prominent examples. Because such trends form a tangled web with social, cultural and political change from which it is difficult to tease isolated threads, simplistic assumptions about relationships between mobility and location are likely to lead to misleading policy prescriptions.[14] However, there can be little doubt about the overall effects: an upward spiral of mobility and dispersal; increasing separation of homes, jobs and services; and travel patterns in which choice rather than proximity (at least for some) has come to dominate. New roads have been demanded to serve new patterns of activity, which in turn have generated their own development pressures, encouraging further changes in land use and more traffic generation (Headicar and Bixby 1992). Purely residential suburbs and villages, out-of-town retail and leisure centres and isolated business parks became the *leitmotivs* of the late twentieth century. So did vehicle pollution, congestion, the dominance of traffic and the decline of traditional centres.

Claims that such trends are unsustainable have been supported by environmental, social and economic arguments. The mutual reinforcement of these different perspectives at a particular moment (in the UK, the early 1990s) was important in the context of the new realism because it allowed the kind of discourse coalition described by Hajer (1995) to organise around the storyline of 'integrated transport'.[15] This coalition was able to exert influence despite differences in the underlying beliefs and values of its members, although, as we shall show, such differences made it unstable. A key driver was recognition of the range and severity of the environmental impacts of transport and land-use trends, combined with the improbability that technical fixes would be sufficient to reduce them permanently to acceptable levels (for discussion, see Banister 1998; Cartledge 1996; Goodwin *et al.* 1991; RCEP 1994, 1997; UK Government 1994a; UKRT 1996a). In the UK, as elsewhere, increases in energy consumption and carbon dioxide emissions from transport seemed almost inevitable.[16] Total emissions of other pollutants, although initially expected to fall as European standards tightened (and to remain below 1990 levels until 2025), were predicted to have resumed an upward trend by 2010 (RCEP 1994). The government itself acknowledged that 'gains could be at risk' if traffic growth continued unchecked and that the downward trend in emissions would not be sufficient in all places to reach air-quality objectives set for 2005 (DETR 1998f: para 2.7). The revised UK air-quality strategy set a less demanding target for particulates – 'the most important air quality challenge' (DETR 2000a: para 260) – on the grounds that an earlier objective (DoE 1997b) was now 'known to be unachievable' (DETR 2000a: para 263).[17] If technical fixes could not cope adequately with pollution, they seemed to have even less to offer in the context of habitat loss, landscape degradation or what for many amounted to a tyranny of traffic in urban and rural areas. In this environmental discourse, the 'green car' would clearly be no panacea for the environmental problems of transport.[18]

An important social dimension was also recognised, more so since the mid-1990s, in line with the changing conceptions of sustainability outlined in Chapter 2. Health, poised between environmental and social dimensions of sustainable development, was a rapidly escalating concern (Fletcher and McMichael 1997; UKRT 1996a). The UK Transport White Paper (DETR 1998f) linked pollutants from road traffic to 24,000 premature deaths annually among vulnerable groups and also noted the concern of the British Medical Association, backed by a growing body of research, that car dependence in children could have serious effects on their physical health and mental development (see, for example, Acheson 1998). Furthermore, developments in land use and transport were severing communities, inhibiting social interaction and polarising society into those with and those without access to a car. The latter group, which remained a substantial one (30 per cent of households in the UK) despite a common misconception that it was somehow residual, was suffering not only from lack of access to a car *per se* but also from deterioration in local services, which could no longer compete with car-oriented facilities. This was becoming a serious new dimension of social deprivation.

Finally, and crucially, there was a powerful eco-modernist component to the new realism, in which certain policies 'derived entirely from arguments of economic efficiency ... turn out to be precisely those policies which contribute most to the protection of the environment' (Goodwin 1993: 268). Trends were seen as unsustainable because congestion imposed heavy costs on business and was detrimental to the efficient functioning of the economy. Even for the Automobile Association (undated: section 3), undeniably a member of the old roads lobby, this was 'bad economics'. And while perspectives on new road construction differed within the transport policy community, 'building our way out of traffic congestion' was generally acknowledged to be neither feasible nor affordable.[19]

Expression of these concerns, significantly spanning a broad political spectrum, became more and more difficult to ignore during the 1990s. Transport was widely acknowledged to be a sector in which 'almost everything' had gone wrong (Pearce 1993: 150). It was possible to agree that trends were unsustainable, at least as a starting point, even in the face of widely divergent conceptions of sustainability: hence the degree of consensus on the need for change, the rapid diffusion of the new realism and the ascendancy of the storyline of 'integrated transport' over that of 'predict and provide'. In many countries, and at different levels of policy making, technological, regulatory, fiscal and institutional measures were proposed in varying degrees, together with investment in improved infrastructure for public transport. Almost invariably, such proposals were accompanied by calls for land-use planning to be better integrated with transport policies and thus to play a prominent role in achieving sustainability – very much an appeal to intersectoralism of the kind that we discussed in Chapter 4. The

House of Lords Select Committee on the European Communities (1994: para 59), having heard evidence from many sections of the transport policy community, noted 'growing agreement' that it was essential for land-use planning to seek to minimise unnecessary movement. Indeed, some elements of the motoring lobby found a convenient scapegoat in the planning system, which, according to the Automobile Association (undated), had probably caused many of the environmental problems associated with transport by permitting development in unsuitable locations. We now turn, therefore, to consider the different facets of this role as it has been perceived and developed, and to review policies that have attempted to promote the integration of land-use and transport planning.

A role for land-use planning?

In one sense, as we have noted, the remit of land-use planning in relation to transport is a traditional one: transport infrastructure is a form of development, and planners have a role in mitigating its often substantial impacts, even if delimiting this remit has itself been a matter of contention. More radically, the planning system might be seen as a forum in which environmental capacities – and thus the 'space' for a new or expanded transport infrastructure – could be delineated. But in a new realist framework, planners found themselves allocated the less familiar role of demand management. At one level, this would involve integrating land-use and transport planning to encourage neglected modes of travel, reversing the ethos of many years in which pedestrians and cyclists had been seen as impediments to the free flow of traffic. On a wider canvas, 'reducing the need to travel' came to be seen, from the early 1990s onwards, as a key task for planners, to be achieved through the instruments of urban design and appropriate location of traffic-generating activities. While the need for co-ordination of development and transport provision has long been recognised, the emphasis on demand management was new, and it resulted in marked shifts of policy in a number of countries.[20]

Transport infrastructure: impacts and capacities

Mitigating the impacts of transport infrastructure on natural and cultural environments has a long history.[21] From 1925 (as a result of the Highways Improvement Act), local authorities could acquire land alongside highways 'for the purpose of planting and amenity', and soon afterwards the Roads Beautifying Association was established, initially as a voluntary body but later as official adviser to the Ministry of Transport (Cullingworth and Nadin 1994). The impacts of traffic growth manifest themselves first in towns and cities, so that already by the mid-1960s the Buchanan Report (Buchanan 1963), in many ways a prescient document, had anticipated severe conflict between valued aspects of the built environment and accommodation of traffic in towns.[22] The virtual cessation of major new road construction in urban areas by the 1970s,[23] although too late to avoid destruction of much that was of value, was effectively a forced acknowledgement of environmental capacity (and financial) constraints (see, for example, Hall 1980). Outside the towns, although the effects of traffic were less immediately acute, conflict between roads and amenity gathered pace as impacts accumulated. Important habitats and landscapes, although taken into consideration in route planning, were not treated as major constraints on the roads programme (Cullingworth and Nadin 1994; Gregory 1974).[24] Annual reports of the Countryside Commission through the 1980s, for example, provide a litany of cases in which roads were constructed in the face of strong environmental objections. The commission found it 'discouraging that departmental and county road planners still accord so little importance to the need to protect designated countryside' (Countryside Commission 1988: 4), and having suffered several 'major setbacks', reacted to the 1989 roads programme with trepidation (Countryside Commission 1990: 3–5). Latterly, greater attention was paid to mitigation and, more controversially, to various forms of 'compensation'. Thus, for example, the much-criticised M3 extension at Twyford Down in the 1990s was accompanied by extensive efforts at habitat relocation and recreation.[25]

In all of this, the town and country planning system had a predominantly reactive role, because the planning of major roads was, until the late 1990s, largely exogenous to it. After the Trunk Roads Act of 1936, highways were planned by a central transport ministry until this role passed to the quasi-independent Highways Agency in 1994. Trunk road schemes were to all intents and purposes superimposed on development plans. Consents were granted by the Secretary of State for Transport, the role of the local authority being relegated to that of statutory consultee, with provision for a public inquiry in the event of objections. In what Dudley and Richardson (1996b: 576) characterise as the 'serene' days of road construction up to the early 1970s, such conflicts as there were centred on amenity, and the most that could normally be achieved to protect important aspects of the natural and cultural environment was some influence on the route – the classic spatial fix – and 'beautifying' by landscaping and design (for a fascinating case study, see Gregory 1974). As a powerful advocacy coalition, highway engineers and the roads lobby had 'managed to structure debate around very few options' (Dudley and Richardson 1996a: 73), so that completion of the trunk road network constituted an aspect of the national transport policy core that was virtually immune to challenge.

During the 1970s, however, conflict over both urban and inter-urban roads mounted as a rival environmental coalition emerged not only to argue about amenity and alternative routes but also to challenge the legitimacy of traffic projections and the need for such developments at all (Tyme 1978).

Public inquiries provided a critical venue in which environmental groups could mount this challenge (Dudley and Richardson 1996a; J. Smith 1994); in doing so, objectors contested the boundaries of the planning system – what could and could not be questioned – illustrating the significance of planning procedures as apertures through which new knowledge, values and beliefs can impinge on public policy. Even so, frustration with the limited remit of public inquiries and, for some, a perceived lack of legitimacy in the formal democratic process led to disruption of some hearings and ensured that the roads programme remained a focus for discontent. By the early 1990s, more than 200 groups were protesting about new roads throughout the UK (Ghazi 1994). In only a small number of cases did such protest result in reappraisal or abandonment of schemes.[26] In some, it led to direct action, including site occupation.

Over the years there were concessions to critics of the roads programme, notably in the way that trunk road schemes were assessed. As conflict intensified, the Department of Transport's system of cost–benefit analysis (CoBA), much criticised for its narrow remit, was supplemented to take account of environmental considerations following the report of the Leitch Committee (DoT 1977)[27], and a manual on environmental appraisal of road schemes was published in 1983 (DoT 1983), effectively anticipating the environmental assessment required by European legislation after 1985. Governments could thus be seen to respond to some of the pressing environmental concerns without reconsidering the main thrust of policy. What was at issue in conflicts over road building was precisely the kind of dilemma that we outlined in Chapter 3. Underlying transport planning seemed to be a premise that however regrettable the loss of natural or cultural assets might be, most were ultimately tradable against the benefits of mobility, which (it was assumed) would be delivered by new roads. For critics of the roads programme, although they may not initially have used this language, such assets constituted critical environmental capital, the protection of which should help to define environmental boundaries for transport policies and programmes. In this context, the 'need' for roads was not immutable, and not only the methodology behind the traffic projections but also the philosophy of making provision to meet demand could be contested. Concepts of environmental capacity, even if not expressed as such, were both constitutive of 'the new realism' and further legitimised by it.

The UK Integrated Transport White Paper promised tighter protection for the natural and built environment. It set out a 'strong presumption' against transport infrastructure that might damage sensitive areas, and it anticipated few cases where 'imperative reasons of overriding public interest' might justify adverse impacts on the integrity of internationally designated sites (DETR 1998f: para 4.200). While avoiding any explicit suggestion of inviolability, the White Paper proposed that concepts of environmental capital being developed by the statutory agencies (see Chapters 3 and 4) would be incorporated into assessment and appraisal 'as appropriate' (*ibid.*: para 4.204). One implication was that the role of planners could move beyond reaction and mitigation to identification of features and functions of natural and cultural environments that should be protected intact, or (when appropriate) traded against the benefits of transport infrastructure only with compensating gains elsewhere.[28]

Significant in this context, because it provides an opportunity to employ the agencies' environmental capital methodology, has been the 'new approach to appraisal' developed at the end of the 1990s, in collaboration with conservation interests, to replace the discredited CoBA (DETR 1998g: Annex B). Ministers had promised that they would 'look at the transport problems which lie behind proposals for road schemes and then ... seek solutions which are environmentally sustainable' (DETR 1997b: 1). Transport options (not just roads) would be assessed against five criteria: economic efficiency, environment, safety, accessibility and integration. Against the various sub-criteria under the 'environment' heading, for example, options would be scored qualitatively, using measures such as 'very serious adverse impact'. Acknowledgement of the need to identify alternative ways of tackling a perceived problem, to assess them against a broad range of criteria, and to do so in a qualitative and relatively transparent way, represents a degree of policy learning, for which planning processes – especially public inquiries – have provided an important forum over several decades. At the time of writing, the new system has been cautiously welcomed by critics of the old (see, for example, English Nature 1999a), but it remains to be fully tested in application. Whether outcomes will be different is an interesting question, not only for transport policy but also for students of policy appraisal (see Rayner 2001). In the context of our discussion in earlier chapters, it is worth noting the resilience of the notion of trade-off and the persistence of the quest for some 'objective' means of assessment that will minimise the need for judgement. For example, it has been suggested that a more explicit weighting system for the different criteria may need to be adopted (English Nature 1999a).

The roads review that followed the White Paper presented an opportunity to apply the new transport philosophy. Its outcome was a 'carefully targeted' £1.4 billion programme (DETR 1998g: 7) of thirty-seven schemes, including a number of by-passes, to be started within seven years (subject to statutory processes). This replaced the inherited £6 billion roads programme of around 150 schemes, already scaled down from over 500 schemes in 1990.[29] At first sight, the review would seem to represent a dramatic reversal of policy, but closer inspection (and subsequent events) suggests that such an interpretation would have been premature. Only thirty-seven schemes were actually withdrawn from the national programme, and many seemed likely to resurface in another guise.[30] By the end of the decade, facing mounting criticism for failing to deliver on transport policy, the government's stance on road building 'as a last resort' was softening. Towards the end of 1999, the Secretary of State asked the Highways Agency to look again at congestion 'hot spots' with a view to accelerating some of the projects in the roads programme (ENDS 1999a).[31] These developments must be seen in a wider political context, to which we return when assessing the prospects for the policies outlined in the White Paper. First, however, we trace changing ideas about the relationship of land-use planning and transport at a smaller scale – that of the neighbourhood – and at a larger and more diffuse one, that of the urban (or larger) region, at which planners have increasingly been expected to manage complex and longer-term interactions.

Reclaiming the streets

In the decades of accommodation, if land-use planning had a role at the local scale it was essentially one of modifying the fabric of the built environment to facilitate the free flow of rapidly growing levels of traffic. Indeed, the main players in the field were not planners but traffic engineers, with surprisingly little communication between these professions, an interesting example of the institutional and cultural barriers to 'integration' that we identified in Chapter 4. The traditional approach to conflict between motorised and non-motorised transport was to subordinate pedestrians and cyclists, and the legacy of this thinking remains imprinted on many urban and rural areas. A later, and not altogether alternative, strategy was to segregate, so that in towns and cities throughout Europe, environmental conditions for those on foot, at least in restricted enclaves, were enhanced by pedestrianisation,[32] and in some countries, notably the Netherlands, segregated cycleways have long been the norm.

The emphasis among the more progressive local authorities later shifted towards a holistic strategy of 'selective accessibility', in which a conscious integration of land-use and transport policies at the local scale attempts to shift the balance of advantage towards environmentally friendly modes of travel. Adopting this philosophy, the historic UK city of York was one of the pioneers of a hierarchy of transport users as the basis for planning, with pedestrians given the highest priority and car-borne commuters and visitors the lowest.[33] This hierarchy was consistent with a broader policy objective of achieving 'development patterns which give people the choice of using more environmentally-friendly means of transport than the car' (City of York 1993: para 8.1). Such patterns were widely held to include networks of safe and relatively direct cycle and footways, compact and mixed urban development related to public transport networks (Bartholomew 1995; Municipality of Schiedam 1991)[34] and integration between environmentally friendly modes (ECMT and OECD 1995; UKRT 1997d).

Inevitably, integrated planning for land use and transport at this scale challenges the hegemony of the private motorist and freight traffic. It has nevertheless gained gradual acceptance and won endorsement in national policies, to some extent led by pressures and responses at local level. In the UK, a spirit of greater co-operation between planners and highway engineers was detectable by the late 1990s, and new guidelines on residential roads and footpaths emphasised permeability of layouts for public transport, cyclists and pedestrians (DETR 1998h).[35] The 1998 Transport White Paper explicitly proposed a reallocation of road space to buses, cyclists and pedestrians, and did so in robust language. It was no longer seen as acceptable for children's freedom to be 'severely curtailed' (DETR 1998f: para 1.14), for people to be 'intimidated' by traffic, or for safety to be improved 'simply [by] discouraging vulnerable groups from venturing on to the roads' (Ibid.: para 3.220). What is significant is that developments at local level, reinforced by new discourses in national policy, have involved a reframing of concepts of freedom and choice, hitherto associated almost exclusively with travel by car. Reinforcing the pressures for change have been the physical problems of pollution and danger manifest at the local scale, campaigns to reduce traffic speeds (Slower Speeds Initiative 1998) and the readiness of some groups to take peaceful direct action to 'reclaim the streets'.

Reducing the need to travel

Interest in the integration of land use and transport planning has not been confined to the local level. Longstanding recognition that land use and transport were interdependent led to calls over several decades for more strategic integration, which could, at least in theory, reduce travel needs, energy consumption and environmental impacts. The rationale was that decisions about location could have a significant impact on travel patterns and that land-use planning, therefore, could become an important instrument of transport policy. By the 1990s, integrated land use and transport planning had come almost to epitomise sustainable development.[36] What is interesting (particularly in the context of institutional learning) is how and why a body of knowledge, some of it quite well established, crossed the threshold of policy significance so that measures aimed at integration were introduced in a number of countries at national or sub-national level. We return to this issue below in reviewing the experience of the UK and the Netherlands, and also that of Portland and Washington County in the USA. First, however, we outline briefly the research findings that provided the legitimation for these ambitious new policies.

There is a history of research in this field dating back at least as far as the 1960s, and a substantial literature reflecting attempts to analyse and model the relationship between land use and transport. The energy crises of the 1970s gave considerable impetus to such work (for reviews, see ECOTEC 1993; Owens 1986; Webster et al. 1988). Modelling, with varying degrees of sophistication, was a favoured tool, but there were also many attempts to demonstrate empirical associations between travel patterns and various attributes of urban form. Given the complexity of the relationships and the essentially technical nature of most of this research (few studies paid much heed to the social and cultural context within which travel and locational decisions were made), it is hardly surprising that it produced few definitive answers. Although it was often possible to demonstrate associations between variables (that between urban density and travel was among the most frequently cited), cause and effect remained hotly disputed (for one particularly heated exchange, see Gordon and Richardson 1989; Newman and Kenworthy 1989), and it proved hazardous to make predictions. So, for example, reducing the physical separation of activities (by increasing densities or the mixing of different land uses) could reasonably be claimed as a necessary condition for reducing the amount of travel by car, but it was most unlikely to be sufficient. More generally, the relative performance of different land-use patterns would depend upon people's propensity to travel, and therefore on a whole host of other considerations. A form that tended to emerge as robust, in that it performed reasonably well under a range of assumptions, was the small to moderate-sized urban area (free-standing or part of a larger conglomerate), with a mix of employment and services accessible by public or non-motorised transport;[37] even so, there were many critics of the simplistic physical determinism that assumed that behaviour could readily be influenced by urban morphology (Owens 1995). In some ways it was easier to say what was inefficient, although one hardly needed a sophisticated computer to show that development in places that were only accessible by car would contribute to the growing volume of road traffic.

The Dutch government was among the first, in the late 1980s, to seek explicitly to integrate land-use and transport planning within the new context of sustainability, with the express aim of influencing the amount and mode of travel (Netherlands Ministry of Housing, Spatial Planning and the Environment 1991b).[38] On the impeccably eco-modernist grounds that 'good accessibility and a high quality environment and living space are vital for the economic and social functioning of the Netherlands' (Netherlands Ministry of Housing, Spatial Planning and the Environment et al. 1994: para 1.3), the 'ABC location policy' related mobility profiles of businesses and amenities, in terms of labour and visitor intensities and vehicle dependence, to accessibility profiles of locations by different forms of transport. Plans, produced jointly for each region by central government, the provinces and municipalities, in consultation with other interested groups, were to classify locations and indicate the types of development to be permitted.[39]

In an interesting example of transnational policy learning, the Dutch experiment exerted significant influence on UK policy in the early 1990s, at just the time when the new philosophy of demand management was calling for novel approaches. A Planning Policy Guidance Note on transport (PPG 13), published jointly by the Departments of Environment and Transport

(1994), was similar in spirit to the ABC policy if less prescriptive in its provisions. It aimed, through a variety of measures, to reduce growth in the length and number of motorised journeys and to encourage more environmentally friendly modes of travel (see Table 5.1). The focus on urban containment and neighbourhood provision of services was far from alien to British postwar planning: the departure in PPG 13 was the explicit attempt to apply a brake to the spiral of mobility and dispersal, in line with the new realism. It was keenly supported by environmental groups.

This emphasis on land-use planning has not been restricted to Europe, although the objectives pose a greater challenge in countries where urban sprawl, dependent on high levels of personal mobility, has been the dominant form of development. An initiative in the American state of Oregon provides an interesting example of policy change in such circumstances. In the late 1980s, opposition to a proposed suburban freeway in the city of Portland, spearheaded by an environmental group, 1000 Friends of Oregon, evolved into a substantial research project focusing on connections between land use, transport and air quality.[40] The primary objective, ultimately successful, was to persuade policy makers to abandon the road scheme and adopt instead a policy emphasising changes in land-use policy and alternative modes of travel. A secondary goal was 'to promote development patterns that reduce land consumption, vehicle trips, and air pollution nationwide'

Table 5.1 UK Planning Policy Guidance on Transport (PPG 13), 1994

Development plans should aim to reduce the need to travel, especially by car, by:

- influencing the location of different types of development relative to transport provision (and vice versa); and

- fostering forms of development which encourage walking, cycling and use of public transport.

Planning and land-use policies should therefore:

- promote development within urban areas, at locations highly accessible by means other than the private car;

- locate major generators of travel demand in existing centres which are highly accessible by means other than the private car;

- strengthen existing local centres and aim to protect and enhance their viability and vitality;

- maintain and improve choice for people to walk, cycle, or use public transport rather than drive between homes and facilities which they need to visit regularly;

- limit parking provision for developments and other on- and off-street parking provision to discourage reliance on the car for work and other journeys where there are effective alternatives.

Source: DoE and DoT 1994: extracts from paras 1.7 and 1.8

(1000 Friends of Oregon 1997: 1; Bartholomew 1995). The project challenged conventional assumptions and created an alternative land-use and transportation plan for Washington County, where, in prevailing conditions, 'most people [had to] use their cars to get to every destination' and where the traditional response to the resulting congestion had been to 'build a new freeway' (1000 Friends of Oregon 1997: 6). This plan was ultimately adopted as part of the region's vision for the future, focusing on mixed-use development, minimum densities and green modes of transport. Action was to be based on three principles (see Table 5.2), which resonate strongly with those of the ABC policy and PPG 13.[41]

Such policies were not adopted purely on a wing and a prayer. Environmental groups and sympathetic policy makers drew upon or commissioned research on land use and transport, then used the results to argue that their preferred approaches could deliver significant benefits. As Flyvbjerg (1998) has shown, those who seek to change the *status quo* and challenge powerful interests have the greatest need for recourse to 'rationality'; in this case, for new policies to be accepted, it had to be 'discovered' that travel demand was related to urban form and density, that out-of-town supermarkets were associated with quite different travel patterns from those at central sites and that employees in remote business parks invariably travelled to work by car. In the UK, a report commissioned jointly by the Departments of Environment and Transport synthesised a great deal of research on such issues and suggested (somewhat heroically) that land-use policies, if combined with a range of transport measures, could lead to a 16 per cent reduction in emissions from transport over a twenty-year period (ECOTEC 1993; see also RCEP 1994). PPG 13 rapidly followed (DoE and DoT 1994). In the Washington County case, the integrated alternative emerged as potentially superior to a 'highways only' option on all key criteria used in the models, including modal split, congestion, job access and air pollution (1000 Friends of Oregon 1997; Bartholomew 1995).

In the case of integrated land-use and transport planning, it was not that

---

Table 5.2 Principles in the LUTRAQ Plan for Washington County, Oregon, USA

- Land use plans should direct higher intensity development to locations well served by transit and should ensure that development is designed for pedestrians, bicyclists, and transit riders, as well as auto drivers.

- The transportation system should serve and reinforce the nature of that development.

- Market strategies should further support the development by correcting some of the current distortions in the pricing of the transportation system and other public facilities.

Source: 1000 Friends of Oregon 1997

analysis over many years proved so conclusive and persuasive as to lead straightforwardly to policy change: uncertainties were always sufficiently great for anyone who disliked the implications of the results to be able to call the assumptions or the methods into question. Rather, as we noted earlier, a political disposition to act could be legitimised, at a particular juncture, by an appeal to 'objective' research. What actually emerged from a great deal of often confusing and contradictory evidence at this point was not a definitive answer but a new *framing* of the interconnectedness of land use and transport. This was both cause and effect of political consternation that the spiral of increasing mobility and dispersal was making car use a matter of necessity rather than choice – that unsustainable patterns of mobility were becoming, almost literally, set in concrete (J. Smith 1994). Providing some support for Kingdon's (1995) 'process streams' approach, the new realist climate of the 1990s enabled a set of solutions to be attached to a (reframed) problem, so that policy change could take place.

At this point, we can conclude not only that transport policies have been changing on a number of fronts but also that the remit of land-use planning in this context has been extended, sometimes consolidating previous experience and raising the profile of land-use issues, but in some cases crossing boundaries into entirely new areas of policy significance. Concern for sustainable development, while it changed few of the fundamentals, undoubtedly furnished an impetus and a frame for a new integration of land-use and transport planning. The last decade of the twentieth century, in particular, saw a marked shift in thinking, in policy discourse and – at least to some extent – in policy practice. But the question remains of whether the developments of the 1990s heralded genuine prospects for more sustainable land-use and transport systems.

Towards sustainability?

One response, as this book goes to press early in 2001, is that it is too soon to tell: policy learning and change can be a lengthy process, with what Weiss (1977: 531) calls 'enlightenment' sometimes taking place over periods of a decade or more. But it is possible to draw some inferences from our overview of policy and to consider potential developments. We might legitimately conclude that more stringent protection for landscapes and habitats, greater priority for green modes, and planning to reduce the need to travel by car have not only become broadly accepted rhetorical objectives but have also been written into policies at different levels of governance. At the level of detail, and of implementation, however, it is less clear either that such policies are sufficient or that they have been (or could be) applied in ways that are likely to bring about lasting change.

For example, it would be difficult to deny that valued environments have come to be seen as more of a constraint on transport infrastructure. But it is a constraint that remains far from absolute. In Britain, even as the new realism was unfolding, a number of highly controversial schemes were given approval,[42] and in the latter part of the 1990s English Nature still regarded transport infrastructure as a threat to protected sites (English Nature 1997a). While the pace of and ambitions for major road construction seemed significantly diminished by the roads review that followed the transport White Paper, forces always pulled in contradictory directions. Powerful lobbies continued to promote investment in roads, albeit within the framework of an integrated transport policy. The CBI (1998: 7–8), for example, pointed out that 'the road network ... will remain the backbone of any successful multi-modal transport system', and the British Roads Federation (1999: 1) that 'in order to keep pace with ... economic growth, the Government must raise the standard of the road network or congestion will become an unbearable burden on both the economy and the environment'. As we have already noted, intensive lobbying quickly resulted in a softening of what had been seen as an 'anti-roads' stance (ENDS 1999a). Further, it seems likely that regional development agencies will advocate resuscitation of at least some road schemes, and that theirs will be a powerful voice within groups charged with taking a wider view of 'transport problems' at sub-regional level. As Goodwin (1999: 6) has argued, we should never underestimate 'the power of an already-designed road scheme'. Such pressures are not confined to the UK. In the Netherlands, the country perhaps most admired by new realists, major transport infrastructure projects caused controversy throughout the 1990s (Bouwer 1994; Ritsema and Asselman 1994), while at a broader European level, roads and other developments, often financed by EU structural funds, have continued to damage prospective Special Areas of Conservation (discussed in more detail in Chapter 6), with the trans-European network being a source of particular consternation (see, for example, Bina *et al.* 1995; WWF 1999; see also Bina 2000).

Meanwhile, in many urban areas facilities for pedestrians and cyclists have slowly improved, and new attitudes have emerged about transport and urban design at the neighbourhood scale, signalling a move away from rigid, car-oriented layouts. Progress on this front has been uneven both between and within different countries, but for the UK as a whole it could hardly be said by the late 1990s that the streets had successfully been reclaimed from traffic. Attempts to integrate land use and transport on a larger scale, exemplified by the Dutch ABC policy and PPG 13 in the UK, have at least helped to ensure that planning and development control take better account of transport – not just traffic – implications than used to be the case. Such changes should not be lightly dismissed. In both the Netherlands and the UK, policies could be judged 'successful' in that the thinking behind them has won broad if not universal acceptance, they have been reasonably well integrated into the development planning process (Netherlands Ministry of Housing, Spatial Planning and the Environment *et al.* 1994; Ove Arup and

Partners 1995, 1999), and they continue to be pursued and refined (see, for example, DETR 1999f). The initiative in Oregon discussed above, as well as having a tangible impact in the abandonment of a particular highway project, influenced regional and state policies on land use and transport and became something of a *cause célèbre*, providing lessons for many other areas in North America (1000 Friends of Oregon 1997).[43]

However, the real test is the extent to which travel behaviour is modified and environmental quality improved, and assessing the effectiveness of policies in these terms is less straightforward. As the Dutch government has acknowledged, 'the effects of location policy on traffic can only be measured in the longer term' and are in any case difficult to isolate (Netherlands Ministry of Housing, Spatial Planning and the Environment *et al* 1994: section 4). While policy makers can point to some specific gains,[44] attempts to use the relatively blunt and long-term instrument of land-use planning to arrest transport trends have encountered numerous obstacles, including difficulties with realising development in the most accessible locations (sometimes because of land contamination), problems in guaranteeing efficient and frequent public transport services (exacerbated in the UK by privatisation and deregulation) and the predictable concern of businesses and local authorities to maintain a competitive edge. Parking standards have been a particular bone of contention. For example, a UK study found that:

> developers tried to retain the level of parking they considered appropriate for their needs, and concentrated on meeting PPG 13 aims by providing support for non car modes, so as to give choice of alternative means of travel rather than to constrain car use.
>
> Ove Arup and Partners 1999: 42

Difficulties have been exacerbated by time lags in the system, which mean that long after policies have been adopted, inappropriate development continues to take place in locations poorly served by public transport (CPRE 1999a; Netherlands Ministry of Housing, Spatial Planning and the Environment *et al.* 1994: Ove Arup and Partners 1995). In the Netherlands, for the period 1991–96, the greatest growth in employment was at 'C' locations, and the development of large peripheral retail centres continued (Netherlands Ministry of Transport, Public Works and Water Management 1999).[45] One study of a medium-sized UK town (Northampton) showed how different factors combine to inhibit effective implementation of PPG 13 (UKRT 1997c). The layout of Northampton (a second-generation new town) presumed high levels of personal mobility; land-use plans still reflected decisions made twenty years earlier; and commercial interests were resistant to car parking restrictions in the town centre, fearing loss of customers to competing towns. In a case with clear parallels to the Newbury example with which we opened this book, a major employer was moving from centrally located offices to a peripheral green field site with expansive parking provision. Although the company was prepared to subsidise bus services for an initial period, such a development seemed clearly to contravene planning policy guidance on transport. Yet the site had outline planning permission for business use, the planning authority was unwilling to confront a major local employer, fearing that the firm would leave the area altogether, and the Regional Office of the (then) Department of the Environment failed to intervene.

An important feature of integrated land-use and transport planning is that localities differ in their inherited built environments, transport systems, economic infrastructure and proximity to competing centres. Not surprisingly, therefore, support for policy change has been geographically uneven, with local government in some places being prepared to take more risks. The Dutch ABC policy encountered least resistance in the Randstad and the Stedenring, where congestion and environmental problems were more apparent (Netherlands Ministry of Housing, Spatial Planning and the Environment *et al.* 1994). In less urbanised areas, or those where public transport was still deemed inadequate, the policy was not so welcome, and some degree of regional differentiation (for example, in parking standards) was necessary for it to gain acceptance. Significant differences between urban and rural areas in responding to PPG 13 have also been found in the UK (Ove Arup and Partners 1999). This illustrates our wider claim that different communities will interpret sustainable development according to their circumstances, and it points to more specific difficulties in identifying sustainable transport policies in rural areas, where both public transport and local facilities have been in decline.

As many critics have pointed out, there is undoubtedly a 'mismatch between the rhetoric of transport policy and the development which is taking place on the ground' (CPRE 1999a: 2); certainly, the amount of travel continues on an upward trend (DETR 1999g). A basic problem is that guidance is 'frequently ignored', or even undermined, by the strategies of other government departments (ETRAC 1999: para 15).[46] At least as important as refining initiatives like PPG 13, therefore, is ensuring that they are interpreted and implemented consistently and are supported by other measures within a national policy framework (CPRE 1999a; ETRAC 1999; Institution of Highways and Transportation 2000; Ove Arup and Partners 1995, 1999). This is crucial, for as we shall argue, neither the difficulties encountered in implementation nor the lack of measurable progress on the ground is surprising when other signals fail to reinforce the message that car use should be reduced, or pull strongly in a different direction. Thus we return to the wider context within which land-use planning has to operate, and in the following section, we ask how radical the shifts in thinking about transport have really been. We draw particularly on experience in the UK, where the fate of policies introduced in the late 1990s provides an interesting test of the durability and resilience of the 'new realism'.

The new realism: planning in context

Having accepted that land-use planning is indispensable in the evolution of a sustainable transport system, the temptation for many commentators has been to load the planning system with ambitious transport and environmental objectives. Yet such expectations are clearly unrealistic: locational policies, for example, cannot single-handedly bring about lower mobility; rather, a propensity to travel less is one of the preconditions for a system in which land-use planning can be an effective instrument of transport policy. The argument for a package including planning alongside fiscal and regulatory measures, as well as investment in environmentally friendly modes of transport, has been well rehearsed; indeed, it has become constitutive of the meaning of 'integration' in transport policy. In the UK, the prospect that land-use planning might become effective in 'complementing and contributing to the success of other measures' seemed enhanced by the approach taken in the Integrated Transport White Paper (DETR 1998f: para 4.156), with its promise of a real shift of emphasis and co-ordination across many policies in the transport arena. But, as we write, the prospects for achieving these objectives seem uncertain for a number of reasons. The White Paper, although in many respects signalling a welcome shift of direction, failed to challenge certain core policy beliefs, and important momentum was lost when the necessary legislation was not immediately forthcoming. Nor is it certain that local authorities will hasten to introduce fiscal restraints on traffic: the conviction of one council leader that this would be 'economic madness' may be far from atypical.[47]

Turning first to the policies set out in the White Paper, it is clear that in crystallising the 'new realism', these represented some radical departures from the approach of the past several decades. Nevertheless, the White Paper remained firmly embedded within a growth paradigm, its ambitions for the most part limited to reducing the *rate* of road traffic growth (para 1.35) rather than the absolute amount of movement. Setting targets against which progress might be measured would be left to local authorities, under the Road Traffic Reduction Act of 1997.[48] Nor should competitiveness be threatened. Indeed, the White Paper stressed that its policies, by reducing congestion and improving the overall efficiency of the transport system, would contribute to this overarching goal, a point also emphasised in the revised draft of PPG 13 (DETR 1999f). This eco-modernist line is best illustrated by the treatment of freight transport: limited modal shift, efficiency gains (such as improving load factors) and technical and spatial fixes (further emissions reductions and time- or place-specific restrictions) were seen as the main ways of achieving 'sustainable distribution' (DETR 1999h), while fundamental questions about the need to move ever-increasing quantities of goods, and related production practices, remained unasked. Yet an efficient transport system serving a competitive economy might still conflict with important environmental objectives.

There were some interests that the government was not prepared to confront. Crucially, in the context of land use and transport, the White Paper did not extend its thinking on parking taxes to out-of-town retail and leisure facilities, while the government went to some lengths to stress that its new policies should not be seen as 'anti-car'. Indeed, in its treatment of the car as a means of mobility (for which there are often plausible substitutes) the White Paper provides an interesting example of 'un-politics' (Crenson 1971), showing how power can be latent in the sense of ensuring that certain issues do not appear on the political agenda at all.[49] Thus it failed to engage with the car as commodity and icon, neglecting profoundly influential aspects of 'car culture' that form an important part of the context for transport policy. Meanwhile, as if in a parallel universe, the marketing of cars continued to emphasise status, acceleration and creature comforts, while, in a parody of the reallocation of road space, the promotion of off-road vehicles opened up new spaces to motorised traffic. Such messages have been demonstrably more powerful than price signals or the kind of 'education' to improve driver attitudes timidly advocated in the White Paper (DETR 1998f: para 3.228). So, we might conclude, the new policies failed to challenge core beliefs regarding growth, competitiveness and consumer sovereignty (certainly to be included among the 'magic symbols' that we discussed in Chapter 3), or openly to confront powerful interests such as retail giants, the motor industry or 'Mondeo man'. It might be argued that to expect all this at once would have been naive; perhaps change must be achieved more gradually, through implementation of the White Paper's other, quietly radical, policies. Even so, political events following its publication suggest that implementation of such policies is likely to remain highly contentious.

That the Transport Bill was afforded relatively low priority in the parliamentary timetable may itself indicate a loss of nerve. While many measures could be enacted immediately, the keystone policies, notably those relating to local authorities' powers to adopt fiscal measures, required primary legislation. This was not forthcoming until late in 2000, nearly two and a half years after publication of the White Paper. This delay allowed potential losers to muster forces and tempted Her Majesty's Opposition to seize upon transport as a party political issue, fracturing the apparent cross-party consensus noted in the introduction to this chapter. What Sabatier (1987: 653) refers to as 'external (system) events' became important too, including a sharp upturn in world oil prices (following a period of decline, which had masked increases in fuel duty), over-capacity in the UK haulage industry and trade liberalisation within the European Union, making UK hauliers feel vulnerable to competition (ENDS 1999b). The 'fuel duty escalator' became a particular focus of discontent after the March 1999 budget, and the government came under intense pressure from road hauliers, the motoring lobby and the Conservatives, who were now committed to abolishing the policy that they had introduced. In November 1999, the Chancellor of the Exchequer announced that in future budgets the escalator would no longer be automatic, and that revenues from any increases in duty above the rate of inflation would be ring-fenced for transport investments. Despite these concessions, a motorist-friendly budget in March 2000[50] and announcement of a £180 billion, ten-year transport plan in July (DETR 2000d),[51] protests about the level of fuel duty escalated. In September 2000, a relatively small number of hauliers, joined by disaffected farmers, blockaded oil refineries, took control of fuel supplies and threatened to bring the country, and the economy, to a halt. They called off their protest only after issuing an ultimatum to the government that it must reduce fuel duty within sixty days. In a pre-budget statement in November 2000, the Chancellor announced a freeze on fuel duties for at least a year, reductions in the duty on ultra-low sulphur fuel[52] and significant concessions on vehicle excise duties (Her Majesty's Treasury 2000).

Although these events loom large as we finalise this manuscript, their longer-term significance is not easy to judge: a predictable (if sometimes dramatic) backlash against the new realism seems unlikely to be the last word. What we can say is that the broadly based 'integrated transport' discourse coalition had fragmented even before the Transport Bill had been introduced in Parliament, leaving divergences of interests and values to emerge into sharper relief. This realignment was accompanied by a widespread sense of frustration with the transport system and dissatisfaction with its rate of improvement – sentiments apparently shared by the Environment, Transport and Regional Affairs Select Committee of the House of Commons (ETRAC 1999), which was sharply critical of the government for the lack of progress. One interpretation of this general mood is that it reflected 'the irritation of impatience, not the hostility of opposition' (Goodwin 1999: 8). An alternative reading would be that rejection (by some) and impatience (more generally) co-existed and reinforced each other. The White Paper had succeeded in raising unrealistic expectations of improvements, while delays in legislation and lack of visible change on the ground allowed opposition to traffic restraint to harden and, in some cases, to be mobilised through a straightforward exercise of (manipulative) power.[53]

Whatever the long-term significance of specific events, it would seem that the timidity of the White Paper on certain fronts, delays and opposition to some of its key measures and the fragility of the consensus on integrated transport point to a number of fundamental difficulties in implementing – even defining – sustainable transport and associated land-use policies. These arise from the association between transport and economic growth (challenged and qualified, but remaining politically potent at every scale), from different framings of the underlying 'problem' concealed (or ignored) by calls for integration, and from divergent ideological perspectives on the extent to which particular demands and preferences should be met.[54] These interrelated issues lie at the root of contradictions in 'sustainable' land-use and transport planning.

Transport and the economy

The relationship between economic activity and the movement of people and goods is both a matter of empirical observation and a source of deeply held normative beliefs. The growth of road traffic has been strongly correlated with economic growth in all developed countries. In the UK, the transport intensity of the economy (passenger or tonne kilometres per unit of GDP) has increased over time (SACTRA 1999) in contrast to energy intensity, for example, which has fallen (RCEP 2000). It might seem, therefore, as a former Secretary of State for Transport told the House of Lords Select Committee on Sustainable Development (1995: para 3.18), that targets for road traffic reduction would 'implicitly [call] into question a broader commitment to economic growth'. Such fears were reinforced by the first report from the Commission for Integrated Transport (CfIT 1999: para 7),[55] which concluded that 'into the medium term, we must continue to regard traffic growth in much of the [UK] as inevitable', a conclusion apparently accepted by the government (DETR 2000e). However, neither suggested that a longer-term decoupling of traffic and economic growth was impossible.

Highly significant in this context was a report by SACTRA (1999) showing that the relationship between income growth and traffic growth was not deterministic. From an extensive review of evidence and empirical studies, the report suggested that:

> ... income growth does have a strong effect on traffic growth, but ... the amount of traffic is also influenced by the price, speed and quality of transport. ... This sensitivity is sufficient to result in a significant degree of variation in how much traffic will arise from any given level of national income. This leads us to conclude that policies intended to change the volume of traffic that will arise from any particular level of economic activity are, in principle, feasible.
>
> *ibid.*: 17

Whether appropriate policies could simply reduce the rate of traffic growth or manage to decouple growth of the economy from growth in traffic is a vital question. CfIT (1999: para 19) concluded that for the rate of traffic growth to be reduced to zero at some future date would require 'intensive application of [policies in] the White Paper, and continued great effort thereafter'. For these reasons, a view that policy should focus on reducing pollution and congestion, rather than the volume of traffic *per se*, has prevailed, conveniently shifting attention towards vehicle performance, traffic management and selective improvements to the road network (DETR 2000e). As we argued earlier, however, there are limits to technical fixes, and they fail to capture some important environmental and social impacts. The amount of traffic therefore continues to be a focus of attention for many groups and, ironically, remains a headline indicator of 'sustainable development' (DETR 1999g).

Framing the problem

A related question is whether increasing transport prices, as one way of restraining traffic growth, might have adverse effects on the economy. Pursuing this line of thought helps us to identify a second fundamental difficulty for radical reform. SACTRA (1999: 18), in addressing the question, made the standard argument that if external costs ('congestion, accidents, pollution and other environmental costs') were included in transport prices, 'the overall marginal costs of a trip to society may be quite different from the direct money costs of car use, or public transport fares, paid by each individual traveller'. Thus:

> The circumstances where reducing traffic levels could contribute usefully to economic performance are, in general, those where transport prices are currently below marginal social costs, primarily because of the existence of external costs of congestion and environmental damage.
>
> *ibid.*

The storyline of 'paying the full social costs' of transport – 'internalising externalities' – arguably acted as an important glue holding together the new realist discourse coalition, being embraced by a wide range of interests from business (for example, CBI 1998) through environmental groups (for example, CPRE 1995) to advisory bodies, including the Royal Commission on Environmental Pollution (1994), CfIT (1999) and SACTRA (1999). But a storyline can be variously interpreted. For some – as in the SACTRA analysis above – it is about making the system more efficient within a welfare economic paradigm. Policies that internalise externalities will 'not only reduce the incidence of such external costs, but also, in doing so, can increase economic welfare' (*ibid.*: 18).[56] The implication is that once prices and costs are better aligned, market forces will produce the correct balance between traffic and environment. But this will not satisfy everyone. Others, while inclined to rally to the cause of making transport users pay their 'full social costs', see using the price mechanism as a means to an end, one that is politically determined with reference, at least in part, to environmental capacities. 'Efficiency', in the conventional economic sense (with its utilitarian underpinnings), is not their ultimate goal.

These underlying differences need not surface when levels of congestion and environmental degradation are, on most assessments, well above some optimum defined in terms of efficiency. In such circumstances, as SACTRA has argued, the goals of economic efficiency and environmental improvement pull in the same direction (although, as we have seen, those who stand to lose may still robustly defend their interests, with some success). But there comes a point 'where transport costs already fully include, or exceed, all internal and external marginal costs, [so that] measures to reduce traffic are likely to entail some sacrifice of economic welfare' (*ibid.*: 18). Those who maintain that environments or human health nevertheless remain insufficiently protected must draw (even if implicitly) on one of the non-utilitarian doctrines that we explored in Chapter 3, moral frameworks in which the right policy might be one whose costs outweigh its benefits.[57] The difference was encapsulated by Brian Mawhinney (1995: 16), who posed 'perhaps the most fundamental question of all' when initiating his 'Great Transport Debate': were we prepared, he asked, 'to curtail our rising economic prosperity to some extent – to any extent – to protect the site of historic significance, the rare flower or the great crested newt and all they represent?' One might add human health to this dilemma, given the conflicts over setting air-quality objectives discussed in this and earlier chapters. In Chapter 3, we showed how different answers to such questions could be associated with divergent conceptions of sustainable development, some permitting trade-offs between environmental and other goals, and others insisting (although not always for the same reasons) that some concept of environmental capacity ought, in all normal circumstances, to be respected. We can now revisit these arguments, and at the same time reconsider the connections between transport, land use and sustainability.

Consumers, citizens and freedoms

A view often expressed is that trends in land use and transport constitute a powerful expression of consumer choice, or (in the terminology used in Chapter 3) 'the wants that people happen to have'.[58] But if we accept that the pursuit of individual preferences in this sphere imposes significant external costs, policies might at least ensure that consumers are informed of the 'true costs' of their actions, through fiscal means and through land-use planning where appropriate. The result would be more efficient, in the terms outlined above, and more sustainable according to conceptions of sustainable development involving trade-offs. Even this much may be difficult, as politicians bruised by conflict over fuel taxes would testify.

Alternative rationales for environmental protection might require policies running more strongly counter to consumer preferences. As we have shown in Chapter 3, satisfactions may be held to have no value if they violate rights or obligations: there are 'boundaries' to systems of human ends (Rawls 1972: 31), even if their precise location remains a matter of intense dispute. In Onora O'Neill's (1996) framework, for example, if policies and behaviours associated with ever-increasing mobility led to systematic or gratuitous injury, justice would require that they be changed (and imperfect obligations might go further). For John O'Neill (1998), the problem is that satisfying the 'wants that people happen to have' is not the best way to contribute to human flourishing. Such arguments, we suggested, often underlie conceptions of sustainable development in which environmental capacities are not to be breached. In the specific context of this chapter, appeals to rights or obligations, or to objective conceptions of the good, may furnish a case for greater restraint on traffic and locational choice than that required for 'efficient' outcomes. For some, such restraints must be coercive: it has been suggested, for example, that 'draconian policies' would be needed to arrest powerful trends (Breheny 1995: 92; see also Goodwin 1993). Others see the loss of 'negative freedom' as acceptable, provided that citizens enjoy the positive freedom to be party to the decisions made,[59] and in the process might produce outcomes that differ substantially from the aggregated preferences of consumers (Sagoff 1988). These are complex and difficult issues. No wonder, then, that policies edging in the direction of 'stronger' conceptions of sustainable transport meet resistance, for they not only challenge powerful interests but also expose profound disagreements about the nature of freedom and choice.

Drawing these arguments together before moving on to more general conclusions, we have shown, using the case of the UK's Integrated Transport Policy, that a challenge to twentieth-century trends in mobility and land use must confront dominant interests, core beliefs and deeply embedded concepts of consumer sovereignty and freedom. Power was latent in the White Paper's failure to take on some of these challenges at all and subsequently manifest in determined opposition to its more radical policy prescriptions. Such resistance need not imply that change is impossible in the longer term, but it does have important implications for those land-use planning policies that have been hailed, almost universally, as a means of promoting a 'more sustainable' transport system. As we have argued above, locational policies should be seen not as an *alternative* to other measures but as providing a means of maintaining accessibility and choice in a future in which mobility is constrained by environmental limits. While a 'sustainable transport policy' has been widely acclaimed, this particular vision of the future, and the conception of sustainability that it implies, has a more restricted following. Land-use policies may therefore continue to work against the grain of a system that in other ways remains conducive to high mobility, in which case they will be ineffective or even counterproductive (for development of this argument, see Owens 1986, 1992).

Conclusions

In the UK, as elsewhere, there were important changes in transport policy discourse during the 1990s, the result of a new framing of problems, political developments such as the growing strength and diversity of the anti-roads lobby, constraints on public expenditure, and international agreements on greenhouse gas emissions. As with many aspects of sustainable development considered in this book, policy practice has shifted too, although it seems fair to conclude that (in the broader picture) changes in behaviour have been barely detectable, traffic levels have grown, pollution trajectories remain a cause for concern, and damage to natural and cultural landscapes has continued. This might be because the 'new realism' needs time to become institutionalised, but a more plausible explanation is that the undoubtedly significant changes of the 1990s have taken place in what Sabatier (1987: 666) calls 'secondary aspects' of the belief systems of policy elites, while important elements of policy cores – the more deeply ingrained beliefs and basic strategies – remain largely impervious to challenge.

What seems to have happened by the end of the 1990s is that a degree of policy learning, in both its discursive and rationalistic forms (reframing of traffic growth as a 'problem' and recognition of the manifest inadequacies of old approaches), combined with particular political events (including the 1997 general election) to make it possible for bold policy statements to be made: 1997 certainly looked like one of Kingdon's (1995) 'critical junctures' for policy change. For a time, the storyline of 'integrated transport' (with an important land-use planning dimension) triumphed over that of 'predict and provide'. But despite the optimism of the White Paper's title (*Better for Everyone*), real change involves winners and losers: as the 'integrated transport' discourse coalition fragmented, different groups began once more to defend their divergent interests and expose their core beliefs. In the more specific sense of co-ordinating land-use and transport policies, 'integration' has not been able fully to overcome 'modes of sectoring' (Degeling 1995: 295), as we might have predicted in Chapter 4.

Changes associated with the 'new realism' have not made commentators optimistic about the prospects for a full decoupling of economic growth from the growth of traffic. If traffic and transport infrastructure continue to impose significant impacts, as seems inevitable, the bleak implication is that environments will be damaged unless stronger conceptions of sustainability prevail. Yet our discussion in this chapter suggests that proponents of policies to constrain traffic growth and its impacts within tightly defined environmental capacities are likely to meet powerful obstacles. In the first place, there are vested interests in high mobility who may be powerful enough to challenge policies that affect them adversely, even when there is evidence that such policies promote the general good. But to suggest that certain environments ought to be protected even if this entails some loss of welfare (as standardly defined) encounters more subtle and pervasive opposition, since it must confront the 'magic symbols' of growth, competitiveness and consumer sovereignty. Within this general framework, land-use policies are unlikely to be able to maintain environmental capital or reduce the level of movement. The power of the 'new realism', broadly associated with the concept of sustainable development, has helped to make transport systems less environmentally harmful than they might otherwise have been. But it has also served to expose, once more, widely divergent conceptions of what it means for patterns of transport – and land use – to be 'sustainable'.

6

PLANNING FOR BIODIVERSITY

Ethics, policies and practice

Biodiversity conservation is an integral part of planning for
sustainable development.

Royal Town Planning Institute (RTPI) 1999: 7

A culture which encourages respect for wildlife and landscapes
is preferable to one that does not.

UK Government 1994b: para 1.28

Introduction

The fate of natural and semi-natural ecosystems is irrevocably bound up
with the use of land. It should therefore be a matter of foremost concern for
all of those involved with land-use change. In this chapter, we explore
formal and conceptual links between land-use planning and nature conservation and consider the influence of growing attention to sustainability.[1]
These connections are of considerable interest for our broader analysis.
Nature conservation policies combine international, national and local
dimensions, making them particularly apposite for attention in the context
of sustainable development. Partly for this reason, their relations with land
use exemplify the 'new remit' for planning, identified in Chapter 2,
requiring acknowledgement of fundamental links between development in
particular localities and environmental change at all scales. The new remit
demands a strategic approach to the conservation and enhancement of biodiversity,[2] able to take account of cumulative impacts even when those of
individual developments are small or appear to have no direct effect on
protected sites; this approach may be frustrated unless biodiversity becomes
a consideration in the development objectives of key economic sectors. Most
interestingly, perhaps, this is a field in which the reconciliatory potential of
sustainability – its claim to 'reduce ... conflict between conservation and
development' (Countryside Commission *et al.* 1993: 9) – seemed to hold out
particular promise. We explore the grounds for such optimism, and the
extent to which this promise has been fulfilled.

As our discussion in Chapter 3 has shown, what is 'natural' is not an unambiguous concept, particularly where nature, culture and economy have co-evolved in long-settled landscapes.[3] While acknowledging this complexity, and remaining sensitive to its policy implications, we focus in our discussion of nature conservation on the (extending) range of species and ecosystems that have become the subject of widespread contemporary concern: many of the latter are semi-natural, and their survival depends upon a dynamic concept of conservation. As elsewhere in the book, we develop the argument primarily with reference to Britain, commenting on the situation in other countries where appropriate. However, it is important to recognise that while institutional arrangements remain specific to particular countries, nature conservation policies have had an increasingly significant international dimension, not least because of the Convention on Biological Diversity, signed by 167 countries at Rio de Janeiro in 1992 (UNCED 1992b). In discussing these policies, and their connections with land-use planning, we concentrate on the strategic level rather than on details of management, regulation and development control, although we touch upon these where they show how policies are being interpreted on the ground. Our aim is to explore key issues at the interface of planning and nature conservation and appropriate frameworks for their treatment within the political process.

We begin by considering the effects of a transition in conservationist thought from a reactive, primarily site-based approach to a more strategic and holistic view based on concepts of biodiversity. Land-use planning continues to be of great importance for site protection, but increasingly it is seen as a forum within which a more proactive and strategic approach to nature conservation can be defined and pursued. And since, in many areas, conservation of what has survived is not enough, it is an approach that must encompass increasing attention to habitat restoration and enhancement. In the UK, these changes are reflected in legislation and in guidance to local planning authorities (from government, statutory agencies and non-governmental organisations), and are set within an increasingly ambitious European policy framework for the protection of biodiversity. At the same time, emerging interpretations of sustainable development, particularly those grounded in concepts of environmental capital and capacity, have been influential in planning and nature conservation policy communities. We examine the novelty, potential and problems of these approaches, confirming our thesis that their initial enthusiastic reception left a number of fundamental questions unanswered.

One of the most enduring of these questions, as we have seen in Chapter 3, is how to defend the 'intuitive ethic' that human beings should protect the 'ongoing, holistic integrity' of nature (Norton 1982: 319), even when conservation would seem to conflict with social and economic considerations. An issue frequently confronted in planning conflicts is the extent to which ecological integrity and preservation of biodiversity might constitute strong prior claims on land use, overriding the case for development. Drawing on our discussion in Chapter 3, we revisit a range of arguments for protective policies, confirming that seemingly incompatible conceptions of what is 'sustainable' are grounded in divergent moral frameworks. However, none furnishes unambiguous guidance for action in every circumstance of conflict, and interpretation of principles is mediated by institutions and power, influencing norms about what constitutes 'harm' and about which social demands are legitimate. It is important, therefore, to look at ways in which concepts of biodiversity and sustainability have been mobilised in land-use planning, examining both the rhetoric and implementation of policies: in doing so, we seek insights into the power of new discourses to challenge dominant paradigms of growth. In the final section, we draw some conclusions about environmental ethics and policy change and identify important challenges for the treatment of valued natural assets in systems that regulate the development and use of land.

Conservation, land and planning

While some protective policies – those prohibiting trade in endangered species, for example – transcend place, institutional structures for nature conservation have traditionally had a clear spatial dimension and have often co-evolved with national arrangements for the regulation of land-use change. The historical, and continuing, emphasis in conservation on site designation and protection (Adams 1996; Bishop *et al.* 1995) has an affinity with planning systems designed to influence the location of economic activities and to deal with site-specific issues in development control. This would certainly describe the situation in the UK, where both planning and nature conservation policies employ longstanding, if not universally successful, mechanisms for the protection of particular sites (Bain *et al.* 1990; O'Sullivan *et al.* 1993; Owens 1993). The core, statutory instrument has been the Site of Special Scientific Interest (SSSI; Areas of Special Scientific Interest (ASSIs) in Northern Ireland), of which there are 6,000 in place covering more than 8 per cent of the land area. Individual sites vary enormously in size: at the top end of the range is the Upper Solway flats and marshes, a wetland complex of 29,950 hectares; at the lower end are sites such as Rockall (0.7 hectares), a tiny island far out in the Atlantic. Sites are designated by the statutory conservation bodies[4] on the basis of established 'scientific' criteria, and, as successive governments have made plain, they 'will continue to be used as the basis for securing the conservation and enhancement of the best sites for wildlife' (UK Government 1994b: para 10.6), including those that have been ascribed international nature conservation value.

Within the growing transnational dimension of nature conservation policy, European legislation has become a major force for change (Winter

1994), and one that has tended to enhance the role of land-use planning. Pre-eminent in this regard is the Habitats Directive,[5] which aims to conserve European species and habitats. Under this Directive, national governments are required to nominate Special Areas of Conservation (SACs), which, along with Special Protection Areas (SPAs) designated under the European 'Birds' Directive,[6] are intended to become part of a coherent Community-wide network of designated sites – 'Natura 2000' – to be subject to stringent protection.[7] Together, the internationally and nationally important sites provide a framework for nature conservation defined outside the planning system but within which this system must operate.[8] Planning authorities are required to consult English Nature (or equivalent statutory bodies) before granting permission for the development of land in, around, or likely to affect any SSSI. In addition, many wildlife areas of less than national significance are given regional or local designations. For example, local authorities can designate Local Nature Reserves (under Section 21 of the National Parks and Access to the Countryside Act 1949), for which demarcation in development plans is the prime mechanism of definition and protection.[9] Many have also identified County Wildlife Sites or Sites of Importance for Nature Conservation.

However, extensive coverage cannot hide the limitations of a site-based approach, even if an understanding of its shortcomings has sometimes been hampered by the patchy, contradictory and politicised quality of monitoring efforts (Adams 1996; Nowell 1991). Despite their protected status, many sites have been lost or damaged as a result of land-use change. 'Development' as defined in town and country planning legislation has accounted for a significant proportion of such damage, and further losses have been negotiated away through ineffective mitigation or compensation measures (Brooke 1996; Cowell 1997; Harrison and Burgess 1994; Whatmore and Boucher 1993). However, the prime culprits have been activities falling outside the remit of the planning system (at least in the UK), most notably farming practices driven by the productivist philosophy of the Common Agricultural Policy (CAP).[10] Some protection has been afforded through the notification procedures and other provisions of the Wildlife and Countryside Act 1981 (and subsequent amendments),[11] environmental assessment requirements,[12] regulations protecting hedgerows[13] and various agri-environment schemes providing incentives for appropriate management rather than harmful practices.[14] However, it is generally acknowledged that such measures have been at best insufficient. Further gaps in regulation include the limited jurisdiction of land-use planning below the low water mark, which has made it difficult to protect sites from the impacts of offshore development such as dredging for aggregates.[15] A more pervasive problem is that sites can decline in nature conservation value, even in the absence of intentional land-use change, if they are not managed in particular ways or if they are subject to more diffuse environmental impacts such as acid deposition, climate change or certain recreational pressures.

For these reasons, it might be argued that the site-based approach frequently fails in its own terms, in the sense of there being inadequate mechanisms to afford protection to the sites that it identifies and designates. The failures have sometimes been spectacular but more often insidious and cumulative. Although official analyses have stressed the growing robustness of site protection and the increase in total area covered (see, for example, English Nature 1997b), prominent failures have potent symbolism, skilfully exploited by those pressing for more stringent controls over activities that cause harm. In 1998, the government consulted on proposals to strengthen protection for SSSIs, including more rigorous tests in development control and the linking of payments to established (rather than prospective) income forgone and to arrangements for positive management (DETR 1998i), measures ultimately incorporated into the Countryside and Rights of Way Act 2000.

However, the limitations of a predominantly site-based approach do not simply arise from the inadequacies of protective legislation. A parallel and in some ways more profound challenge has emerged from changes in ecological thinking, affecting ideas about sources of value in nature and how it might be conserved. As Bishop *et al.* (1995: 295) observe, those asking questions 'are not only the traditional opponents of protective legislation ... but also some in conservation circles'. These changes include a growing recognition of connectedness and a consequent reconceptualisation of sites as components in wider planning hierarchies and ecological networks. In this more holistic view, nature conservation demands strategic environmental planning 'based less on traditional defensive approaches and more on creative and adaptive techniques' (*ibid.*: 298). Furthermore, the necessary objectives and policies must often transcend administrative and national boundaries. Hence the emphasis in the Habitats Directive on the formation of a coherent European ecological network and on the management of features that 'because of their linear and continuous structure or their function as stepping stones, are essential for migration, dispersal and genetic exchange' (DoE 1994a: para 16). In such a framework, site protection remains necessary, but seriously insufficient.

In Britain, too, a long-established national policy framework for nature conservation has been adapted to reflect these changing perspectives and priorities. One effect – not least because of the need to implement European legislation – has been new recognition and enhancement of the role of land-use planning and the development of a closer affinity between land-use planning and nature conservation policy communities. The Planning and Compensation Act 1991 requires that development plans include policies in respect of the conservation of the natural beauty and amenity of the land,

and planning policy guidance emphasises the importance of decisions made by local planning authorities for the 'sound stewardship' of wildlife and natural features. The guidance also stresses that protection of wildlife 'depends on the wise management of the nation's land resource *as a whole*' (*ibid.*: para 4, emphasis added), mapping the global language of biodiversity across a far wider terrain than that of protected areas:

> Statutory and non-statutory sites, together with countryside features which provide wildlife corridors, links or stepping stones from one habitat to another, all help to form a network necessary to ensure the maintenance of the current range and diversity of our flora, fauna, geological and landform features and the survival of important species.
>
> *ibid.*: para 15

Indirect, systemic threats to wildlife habitats are also given recognition in the extension of 'consultation areas' around national and international sites, in which planning authorities are required to consult the statutory conservation agencies about possible effects from development proposals (*ibid.*: paras 30–2).

Contributing to a more holistic policy context has been a proliferation of other frameworks, helping to extend the 'material consideration' of nature conservation in planning decisions across a broader sweep of the landscape. The British government's response to the Biodiversity Convention, the UK Action Plan for Biodiversity (UK Government 1994b), which sets out an ambitious programme of work in the context of Britain's sustainability strategy, acknowledges that many plants and animals 'are not generally amenable to site based conservation initiatives but instead require the retention of such features of the wider countryside as hedges and copses, ponds and flushes' (para 4.25). Action plans for species and habitats are an important constituent of the strategy, for which site protection is just one part. During the 1990s, the thinking of statutory advisory bodies also moved beyond site and protected area boundaries. English Nature created a map of 'Natural Areas' as a framework for 'characterising features, species and habitats and taking an integrated approach to their conservation needs' (English Nature 1993a: 2), and the Countryside Commission attempted a comprehensive analysis and mapping of 'countryside character' (English Nature and the Countryside Commission undated). These efforts – interesting examples of the desire (noted in Chapter 4) to document and raise the profile of environmental qualities – have been combined in a map of England that 'reflects the natural and cultural dimensions of the landscape' (*ibid.*: 1).[16] Similarly, Scottish Natural Heritage has instigated 'natural heritage zones' (SNH 1997). As we noted earlier, however, attention is not restricted to what has remained intact. The desire for a more proactive approach is reflected in programmes of restoration, involving reinstatement of habitats deemed 'natural' or appropriate to particular contexts, with associated management practices (Adams 1996; Phillips 1996). Ambitious plans for Wicken Fen and the surrounding area in Cambridgeshire provide an interesting example (Friday and Coulston 1999). Such schemes often involve a wide range of institutional partners, but planning obligations and conditions offer one mechanism for securing the necessary land and financial resources.

Finally, in the context of changes in ecological thinking, effects (on ambitions, at least) can also be detected at local level. From a rather conventional and uneven coverage of nature conservation issues in structure plans during the 1980s (Bain *et al.* 1990), a broader approach has developed, incorporating a growing number of non-statutory nature conservation strategies and, latterly, Local Biodiversity Action Plans. In line with the new participatory ethos discussed in Chapter 4, the object of these plans is to engage the public and stakeholders in developing local strategies to achieve national biodiversity targets (Scottish Biodiversity Group 1997; UK Local Issues Advisory Group 1997a–e). Such frameworks extend identification of important natural assets beyond sites to more extensive habitats and the locally characteristic, promoting policies for their conservation that may subsequently be reflected in statutory development plans. In turn, these statutory plans may help to deliver biodiversity targets (UK Local Issues Advisory Group 1997c). The growing use of environmental audit-type procedures has also helped to raise the profile of natural assets, not only in (statutory) landuse planning but also in Local Environment Agency Plans, agri-environment schemes and the forward planning of utilities such as electricity and water (Peterborough Environment City Trust 1995).

In a number of ways then, planning policies for nature conservation have undergone a significant transition in which the broadening of a predominantly site-based perspective has been intertwined with the rise to prominence of the concept of sustainable development. While much effort has been devoted to repositioning protected areas as 'part of the jigsaw of sustainability' (Bishop *et al.* 1995: 304), a more holistic, multi-scale approach has not necessarily made planning practice any easier. Certainly, the area over which nature conservation might be a consideration has been extended, blurring the dualistic thinking that separated spaces for development from spaces for protection (Whatmore and Boucher 1993). But, at the same time, a broadening of the reach of conservation interests has expanded the range of situations in which environmental, social and economic concerns might come into conflict, so that policy dilemmas become almost omnipresent. Moreover, a more systemic appreciation of nature and its interplay with cultural processes has done little, of itself, to increase the strength of protection or the weight given to conservation interests in planning processes. In this domain, the concept of sustainable development seems to have been more directly influential.

Nature as 'environmental capital'

At first sight, the concepts of critical and constant environmental capital, introduced in Chapter 3, seem to have a powerful affinity with strategies for nature conservation. Both long-established and more recent approaches to conservation have, in a sense, involved the identification and safeguarding of elements of 'environmental capital'. Indeed, many conservationists, as well as planners, embraced the ideas emerging in the 1990s with enthusiasm, seeing them as helpful in promoting environment-led conceptions of sustainability in land-use planning. The Royal Society for the Protection of Birds (RSPB 1993: 18), for example, one of the largest non-governmental organisations promoting nature conservation in Britain, saw the planning system as a framework 'for weighing up the merits of alternatives, defining critical natural capital and sustainability constraints, and exploring mitigation in respect of non-critical "tradable" assets' (see also Jacobs 1993). At about the same time, the statutory conservation bodies produced statements supportive of these ideas (see, for example, Countryside Commission 1993; Countryside Commission, English Heritage and English Nature 1993; English Nature 1993b). English Nature's pursuit of environmental sustainability was to involve opposing development and land use 'which adversely and irreversibly affects critical natural capital' and encouraging, in tandem with development, 'the maintenance and enhancement of natural features to ensure an increasing level of net natural assets' (*ibid.*: 2).

Such statements are indicative of the widespread acceptance of a distinction between critical environmental capital and other assets, the former to be protected intact within a wider programme for maintaining the value of some environmental capital stock at a non-declining level. From a land-use planning perspective, delineation of wildlife habitats, partly exogenous to and partly within the planning process, could form part of the environmental baseline for policy formulation and would be constitutive of the environmental (and other) objectives of the development plan; it might also provide an input to strategic environmental assessment. Such thinking is well illustrated by English Nature's early work on sustainability (David Tyldesley and Associates 1994), which set identification of critical and constant natural assets within the survey stage of a linear rational model of the planning process. Critical natural assets were defined both as sites, notably SSSIs, and as thresholds, such as population levels below which a particular species would not be viable or levels of water reserves or air quality required to sustain critical habitats. For these assets, the aim would be 'no loss, damage or decline' (*ibid.*: 29). We suggested in Chapter 3 that the concept of criticality seemed to promise a rationale for strong protection, one that was particularly welcome in a policy community that saw itself as fighting a rearguard action against relentless development pressures and land-use change. Similarly, a broad emphasis on maintaining overall levels of environmental capital legitimised the extension of conservation interests across the wider countryside, beyond special sites, to encompass both ecological quality and landscape character. All this, it seemed, exuded an admirable clarity and parsimony with the policy rules that it generated.

Predictably, however, application of such models in a specific context – nature conservation – was fraught with problems. From the outset, it was clear that the static metaphor of 'capital' sits uncomfortably with the dynamics of natural processes, and it was ironic that enthusiasm to inscribe environmental capital into development plans, with their spatial emphasis, coincided with growing recognition of the limitations of site-based conservation. Even if many important sites would in practice be identified as critical, sustaining the 'quality and value of the natural environment' would have to amount to more than protecting 'core areas' (Shepherd and Gillespie 1996: 2; Gillespie and Shepherd 1995). More substantially, new understandings about the non-equilibrial nature of ecosystems imply that 'recognising a nature "out there" to respect is much more complicated than it may seem' (Demerrit 1994: 26), and providing space for natural *processes* to 'function outside human planning' (Adams 1996: 173) may be at least as important as the protection of specific entities. Some see maintenance of the surviving assemblage of species and habitats as inadequate for different reasons, arguing a moral or scientific case for reversing historical losses and honouring a 'natural debt';[17] in this spirit, governments and conservation bodies have set targets for species and habitats drawn from the idea of 'safe minimum standards' and from notions of what is indigenous. Lowland wet grassland (English Nature *et al.* 1998), heathlands, forests and flood plains are just some of the habitats subject to enhancement and conservation plans.

As thinking has developed and has been tested by the necessary but exacting process of implementation, the concept of a stock of environmental capital, and its purchase on policy design, have become more nuanced. In some contexts, in certain estuary strategies, for example, the model has been eschewed altogether (Jemmett 1995), being seen as insufficiently dynamic and unhelpful when attempting to establish an inclusive strategic process.[18] Elsewhere, considerable effort has been devoted to adapting definitions of environmental capital to soften its spatial dimension, simultaneously making its interpretation in land-use planning more complex. Thus English Nature has recognised that criteria for defining natural capital in the uplands of England, where 'large continuous tracts of semi-natural habitat … are an integral part of farming and other land management systems' (Shepherd and Gillespie 1996: 2), will differ from those suitable for intensively farmed areas, where land with significant wildlife interest is thinly dispersed and fragmented (see also Gillespie and Shepherd 1995).[19] But refinements have also been driven by pragmatic, political considerations. A classification that seemed initially to promise a coherent framework for *stronger* protection looked, on closer inspection, as if it might have the opposite effect. If criticality implied near inviolability, it did not require great prescience to see that few sites were likely to be deemed critical: others might be seen, by implication, as dispensable. Hard-won policies for site conservation with statutory backing risked being undermined, and the dilemma inherent in all systems of hierarchical protection would be exacerbated. English Nature therefore soon distanced itself from suggestions that concepts of environmental capital could simply be mapped onto existing hierarchies of designation:

> For the present, there should be no link made between the designation of a site and the identification of environmental features as [critical natural capital] ... Identifying [critical natural capital] is a step towards implementing environmental sustainability by ensuring the maintenance of biodiversity. Many designation systems are aimed at conserving a representative range of our biota and are not necessarily planned in such a way that they ensure sustained biodiversity maintenance'
>
> Gillespie and Shepherd 1995: 9

Similar thinking informed English Nature's collaboration with the other English statutory conservation agencies in their joint project on environmental capital (CAG and Land Use Consultants 1997; see Chapter 3). In this, the emphasis when assessing importance and replaceability has been on the functions or services provided by environments, rather than on whole entities such as forests or rivers. This approach circumvents some of the difficulties associated with the environmental capital metaphor but retains or introduces others: its language (at least) remains instrumental; there is a danger of reductionism in focusing on 'services'; and, as we argued earlier, this shift of emphasis cannot displace the centrality of judgements and moral choices. However, that the approach holds some attraction for decision makers has been signalled by references to it in important policy statements such as the Integrated Transport White Paper (DETR 1998f).

In short, the quest for a robust rationale for nature conservation has not ended with the discourse of 'maintaining environmental capital', although protecting biodiversity might have been reframed as part of this wider project. In our discussion of sustainability and environmental capital in Chapter 3, we argued that competing definitions and prescriptions not only (transparently) reflect different interests but could also be grounded in fundamentally different sets of values and beliefs. It should come as no surprise, therefore, that interpreting 'nature' as environmental capital does not of itself provide a trump card for conservation but leads us back to the kinds of ethical and political choices identified in our earlier discussion: what is it in nature that has value and merits protection, and when and why should such values take precedence over other claims? To take but one instance, the question of what constitutes, and what habitats might support, a sustainable population of bitterns (an endangered marshland bird species for which there are targets in *Biodiversity: The UK Action Plan* (UK Government 1994b)), can be informed by ecological science, but in the light of considerable uncertainty, choosing appropriate actions must depend on non-scientific judgements about precaution. Further, and more fundamentally, while ecology may be able to suggest how targets might be achieved, science alone cannot tell us why it is right or good to promote the conditions in which bitterns could thrive in the UK.

In defence of nature

Divergent views about such issues underlie and intensify conflicts between development and nature conservation and in 'hard cases' may point towards different and incompatible outcomes. Damaging the natural world often feels wrong, for example, even when development would clearly deliver net benefits to society. Defending protection in such circumstances implies a moral framework in which the scope for trade-off is restricted, perhaps by taking obligations or rights as the fundamental ethical categories. We explored one such framework in Chapter 3, showing how an obligation to reject the principle of injury could lead to an account of environmental justice as 'a matter of sustaining life and lives without avoidable damage to the reproductive and regenerative powers of the natural world' (O. O'Neill 1996: 177, 1997). This would mean adopting demanding precautionary policies in sectors such as agriculture and transport and, significantly for land-use planning, would 'spell out some constraints on what may be done in a given time and place, with its actual resources and population, if agents are not to act on the environment in ways that will or may injure systematically or gratuitously' (O. O'Neill 1997: 137). These constraints may be tighter than those implied by any calculation of costs and benefits: certainly, Onora O'Neill (1996: 177) suggests that by the standards of her account, 'many commonplace activities and policies are environmentally unjust'.[20]

Those who see conservation as a matter of justice, or obligation, are unlikely to be persuaded by the argument (however well supported by technical evidence) that some proposed development would maximise well-being in terms of preference satisfaction. Conflict may be even more intense when different beliefs about the moral standing of the non-human world are involved. In Onora O'Neill's account, while the case for conservation does not depend on maximising utility, it is grounded in a (known or potential) commonality of interests between humans and the non-human world: because 'injury' may be caused by damage to the 'material basis for human life and livelihood' (*ibid.*: 168), avoiding it may furnish protection for nature as well as just treatment for other humans. But if individual living things, species or ecosystems have 'goods of their own' and intrinsic (or non-instrumental) value, the degree of protection afforded by this framework may still be insufficient.[21] As Banner (1999: 170) has argued, 'even if the interests of humans and of the natural environment are not always simply opposed, there are surely cases where they are not simply compatible'; and it may not be axiomatic that human interests should always take precedence. Many environmentalists reject what they see as an overly anthropocentric bias in decision making, believing that non-humans should be afforded ethical standing in their own right. An extensive body of work in environmental ethics seeks to underpin this case. Johnson (1991: 118), for example, argues that (human) agents should give '*due* respect to all the interests of all beings that have interests, in proportion to their interests', and Taylor (1986) proposes a guiding principle that non-basic human needs should not be permitted to outweigh the basic needs of non-humans.[22] For others, the moral injunction to protect and cherish nature is based on different but equally controversial premises: that preferences for conservation are simply ones that it is 'right for people to have' (Banner 1999: 172), values that we have a duty to cultivate. Thus the flourishing of many living things *ought* to be promoted 'because they are constitutive of our own flourishing ... care for the natural world for its own sake is a part of the best life for humans' (J. O'Neill 1993: 24).[23] Yet, as we have seen in Chapter 3, objective conceptions of the good run counter not only to cherished liberal principles but also to the driving forces of the economic system. Whether, in a liberal society, environmental values are widely shared remains in dispute (for different perspectives, see Hargrove 1992; Lowe and Goyder 1983; Sagoff 1988). In practice, the view that protecting natural environments is constitutive of human flourishing often conflicts with what proponents of development believe to be good.

These issues will continue to be the subject of much energetic debate. Our purpose here has been to emphasise, in the context of nature conservation, that deep differences of value lie close beneath the surface of conflicts over the use of land, and that appeals to a singular concept of sustainable development cannot hope to resolve them. Of course there is common ground: authors starting from very different premises can often agree in principle that we should not destroy species or ecosystems without good cause, and that 'when we do, we ought to proceed only with moral consciousness, and with caution and restraint' (Johnson 1991: 200). Even within a given framework, interests or duties may conflict, and none of the approaches outlined precludes all change or intervention in the natural world. However, the most testing cases are those in which different underlying values have sharply divergent implications, and these, we suggest, are reflected in the intensity of conflicts over legislation (in which principles become institutionalised) and over specific claims on land use. Views on which beings should have ethical standing, for example, 'range widely, even wildly', and one suspects that the 'interminable and inconclusive controversy' (O. O'Neill 1996: 94) on this subject is repeatedly played out in equally inconclusive wrangles of the 'jobs versus nature' variety. Similarly, we see an enduring tension between developers seeking to satisfy 'the wants that people happen to have' and those who believe that a desire to cherish nature is objectively 'better' than many other preferences. Such differences are reflected, but not reconciled, in the treatment of 'environmental capital' and alternative conceptions of sustainable development.

Finally, we must acknowledge that, for the most part, societies approach the necessary choices not directly from basic principles in each case but through a range of institutions that constrain the available options. National and international legal frameworks for nature conservation and planning can be seen as ways of formalising ideas about what is considerable and in what circumstances. We now turn, therefore, to assess the significance of under-lying beliefs and values in the real world of political bargaining and power, exploring both the rhetoric of conservation policy and actual planning outcomes.

Interpreting protection

Policies and legislation

Policies for nature conservation and protection of biodiversity rarely draw explicitly upon a single ethical framework: more typically, an ethical basis – and often more than one – is implicit in policy objectives and in the ways in which they are rationalised and defended. There has been a tendency to defend conservation (and sustainable development) in instrumental and util-itarian terms: biodiversity delivers direct and indirect benefits to humans, which, with suitable allowances for uncertainty, will often (but will not invariably) outweigh the costs of protection. As the UK's *Biodiversity Action Plan* argues, the conservation of biodiversity is not without cost: 'We must therefore be clear why biodiversity matters so that a balanced view can be formed' (UK Government 1994b: para 1.14). In this plan, as in many other policy statements, material and instrumental values are prominent, linking concern for biodiversity to survival and health, natural protective functions such as flood plains, and a variety of aesthetic and economic considerations: for example, the possible cancer-curing properties of the Pacific yew are stressed in the introductory chapter. Similarly, local biodiversity action plans are seen to be 'as much about recognition of the potential economic impor-tance of the natural environment as ... about improving the quality of life' (Scottish Office 1997: 6). In this vein, concerns about constraints on devel-opment are countered by claims about employment and income benefits associated with conservation (see, for example, SNH 1998), an interesting manifestation of the 'analytical arms race' that we discussed in Chapter 4.

Although such values have long been stressed in defence of conservation (see, for example, Samuel Hays' informative 1989 history of the American

environmental movement) and have been given renewed impetus by instrumental discourses of sustainable development, it remains difficult to
characterise the range and intent of protective policies purely in such terms.
Much national and international legislation (at least on paper) confers strong
protection upon species and habitats beyond what could plausibly be justified in terms of optimising human welfare, even if prudence is stretched to
the limit. The US Endangered Species Act comes close to creating an inviolable status for designated species (Baden and Geddes 2000; Gerrard 1994),
while, in a more spatially specific sense, Norwegian legislation rules out the
possibility of hydro-power on a number of undeveloped river systems. The
first UK strategy for sustainable development acknowledged that there were
limits to trade-off: 'Sometimes environmental costs have to be accepted as
the price of economic development, but on other occasions a site, or an
ecosystem ... has to be regarded as *so valuable* that it should be protected
from exploitation' (UK Government 1994a: para 3.15, emphasis added).
And it would surely be challenging to defend all of the targets adopted in
the *Biodiversity Action Plan* (UK Government 1994b) on grounds of utility
and precaution. As a final, and important, example, the European Habitats
Directive severely circumscribes the potential to 'balance' or trade off the
integrity of Special Areas of Conservation against economic and social benefits. Where development threatens a site hosting a priority habitat or
species, only human health and safety, or 'beneficial consequences of primary
importance for the environment', are permissible considerations, unless there
are other 'imperative reasons of overriding public interest' to proceed with
the development.[24] For many, as we have seen, the metaphor of critical environmental capital encapsulates the status of aspects of the natural world
requiring such rigorous protection.

Non-instrumental and non-utilitarian rationales for nature conservation
are therefore implicit in a range of policies and are sometimes more explicit
in the reasoning, when concepts of intrinsic value in non-human nature are
connected with human flourishing and with various sets of responsibilities
and obligations. The UK *Biodiversity Action Plan* answered its own question
as to why we should conserve biodiversity not only with arguments about
utility and precaution, but also by acknowledging an obligation to safeguard
the non-human world ('with ... dominion comes responsibility'), as well as a
duty to 'hand on to the next generation an environment no less rich than the
one we ourselves inherited' (*ibid.*: para 1.28). In its bold assertion, quoted at
the beginning of this chapter, that a 'culture which encourages respect for
wildlife and landscapes is preferable to one that does not' (*ibid.*), there is a
distinct hint of an objective sense of the good: species and habitats 'enrich
our lives' (*ibid.*: para 1.27), so that 'we shall suffer both economic *and spiritual* loss' (*ibid.*: para 1.31, emphasis added) if we fail to protect biodiversity.
Such statements imply that protection of biodiversity may be right even
where its costs are greater than its benefits, although it does not follow that

it is always right, whatever the costs.[25] The caveats in strong protective legislation might be interpreted as an acknowledgement of this, by specifying which kinds of consideration could override a duty to protect a particular natural asset, in effect moving towards 'stratified levels of moral pertinence and policy relevance' (Hodgson 1997: 57).

Two points are of particular significance here. One is that while policies for nature conservation are frequently justified in material, instrumental and utilitarian terms, there are also important elements implying that not everything can or should be weighed against everything else in a quest to maximise some measure of human well-being. The second is that specific questions of what is of value and what is right tend, as we suggested in Chapter 3, to be subsumed into a spatial hierarchy of significance when structuring permissible trade-offs. UK planning policy guidance requires local planning authorities to 'have regard to the relative significance of international, national, local and informal designations when considering the weight to be attached to nature conservation interests' (DoE 1994a: para 4), and advice on good practice reinforces this hierarchy (for model policies, see RTPI 1999: 27). Only for *internationally* important sites does legislation afford the most rigorous tests, in which the scope for trade-off is strictly limited. For others, it seems that all considerations can be put into the balance, although for SSSIs at least, it is assumed that nature conservation should weigh heavily.[26] Not surprisingly, such hierarchies also permeate environmental capital methodologies, with a prime consideration being *at what scale* a habitat or attribute of the environment can be said to matter (see Cole 1997: 4). The principle envisaged by some commentators (for example, Bain *et al.* 1990; Countryside Commission *et al.* 1993) – that only pro-development interests significant at at least the same scale as that of the designation should be permissible planning considerations – has not been applied comprehensively, although such thinking clearly influences decisions and is implicit in references to 'the national interest'. Even so, as we have argued in earlier chapters, the spatial level of significance does not tell us all we need to know about the 'rightness' of development or conservation in any given case.

What this overview suggests is that policies for nature conservation cannot be accounted for in purely human instrumental terms. Nor are they always grounded in utilitarian principles; they already indicate a degree of respect for some form of intrinsic value in non-human nature and often reflect a sense of obligation to protect biodiversity ('for its own sake' or for present or future humans). If anything, policies have been strengthened both nationally and internationally as concern for sustainable development has grown. However, what really matters is how such policies are interpreted and implemented on the ground, and it is here that systems of land-use planning become vital. In the planning process, the values that 'disport themselves with happy plurality' in general policy statements must indeed be brought 'sharply into line' (Foster 1997: 235), and vague generalities about sustainability have to be translated into hard decisions about the use and development of land.

Conservation in practice

Policy rhetoric is of considerable significance for planning if, as we have argued above, decisions about land use are increasingly important in the delivery of biodiversity objectives. But we need to examine how these challenges are being taken up in the processes of planning and development control and the extent to which particular plans and decisions accord with the spirit of protective policies and legislation. Throughout the 1990s, as increasing emphasis was placed on the broad consensual goals of sustainability and conserving biodiversity, nature conservation acquired a higher profile in land-use planning, and vice versa. In the UK, planning bodies became more proactive in terms of information and auditing, policy formulation, and local strategies or action plans, sometimes combining these activities in innovative ways. Some experimented with, or adopted, the environmental capital approach. Reviews and reports of 'good practice' proliferated, indicating (as with sustainable development more generally) widespread interest in biodiversity objectives, if somewhat patchy progress and uncertainty about exactly what to do (see, for example, Brooke 1996; Byron 2000; CAG 1998; DETR 1998c; RSPB 1999; RTPI 1999; UK Local Issues Advisory Group 1997a–e). At the forefront were authorities like Peterborough City Council, which, in partnership with others, attempted a systematic evaluation of local environmental capital as the basis for a planning strategy and adopted it as supplementary planning guidance (English Nature 1995; Peterborough Environment City Trust 1995).[27] At a larger scale, the South West Regional Planning Conference participated in a biodiversity audit for the region, which formed the basis for a regional biodiversity action plan (Cordrey 1997).[28] Local authorities also became more energetic in the designation of statutory and non-statutory sites for nature conservation and in participating in local biodiversity action plans: for example, there was a sharp increase in the 1990s in declarations of local nature reserves, especially in urban and semi-urban areas,[29] and by late 1998 there were well over 100 local biodiversity action plans throughout the UK. This activity has grown alongside recognition of the value of direct contact between people and nature, even in sites of relatively low scientifically defined conservation value (Adams 1996; Box and Harrison 1993; UK MAB Committee 1998) and projects such as parish mapping, which encourage people's involvement by promoting what is characteristic or distinctive in particular localities (Clifford 1997; LaViolette and McIntosh 1997; English Nature 1998a). Through such engagement, it might be argued, appreciation of the natural world is fostered and support for protective planning policies deepened (Faulkner 1999).

Site protection

Trends in what we might call the 'decade of sustainability' seemed to bode well, therefore, for making nature conservation an integral consideration in land-use planning. At least as far as conventional site protection was concerned, the planning system was becoming increasingly effective, allowing English Nature (1997b: 11) to claim that 'loss and damage to SSSIs from inappropriate development is now at a very low level'. The threats posed by the roads programme, as we saw in Chapter 5, had (at least temporarily) receded,[30] and European legislation was beginning to give the planning system more bite. A significant precedent was set in the 'Leybucht–Dykes' case in 1991, when the European Court of Justice ruled that derogations of the duty to protect an SPA might be permitted only if they entailed 'a general interest superior to the ecological objective' envisaged by Article 4(4) of the Birds Directive.[31] The reduction of risks to human life was held to satisfy such a criterion, but not economic or recreational interests (Miller 1999b). Local authorities began to make effective use of the legislation: Kent County Council, for example, invoked the Habitats Regulations to revoke an old planning consent for land reclamation affecting the Barksore Marshes, part of the Medway Estuary and Marshes SPA, an order confirmed by the Secretary of State after a public inquiry into objections from the developer, the land owner and the port.[32] Economic interests were not invariably holding sway, and SSSIs were further safeguarded in the Countryside and Rights of Way Act 2000.

There is some evidence, then, that an obligation to protect biodiversity was being taken seriously, at least where it was interpreted through the processes of development control. As many critics pointed out, however, there was no cause for complacency, even in the traditional sphere of site protection (see, for example, Lawson 1997; Le-Las 1999; WWF 1999; WWF UK 1997). While the rate of destruction of and damage to SSSIs had declined, and total coverage increased,[33] losses to development persisted,[34] and many sites still suffered from neglect, lack of positive management or unsustainable agricultural practices.[35] Others were threatened by water abstraction or minerals extraction, for which there were numerous outstanding consents. Many of these pressures lay wholly or substantially beyond the reach of the planning system, a sharp reminder of its limitations in influencing ecological change. In the wider countryside, enjoying no special protection, the situation was seen as one of 'relentless decline' (CPRE 2000a: para 13), and it might be one in which land-use planning has even less purchase. In comparison with the long-established practice of site protection, a holistic approach to nature conservation, perhaps including policies of 'no net loss' (English Nature 1998b: 6), has proved more difficult to integrate with conventional planning processes: as one practitioner observed, 'sites are "easier" than species' (RSPB 1999: 32). For similar reasons, enhancement of environmental capital – for example by adopting biodiversity targets and policies conducive to achieving them – is likely to be even more challenging in a planning system unaccustomed to being proactive in this sphere. 'Integration', as we suggested in Chapter 4, may be more complex and difficult in reality than the rhetoric implies: by 2000, despite exhortation and increased activity, fewer than 20 per cent of English planning authorities had used biodiversity action plans to inform development plan policy (English Nature 2000), and only a small proportion of development plans actually contained effective policies to encourage positive management in the wider countryside (David Tyldesley and Associates 2000).

Even the relative successes in the area of site protection have some perturbing implications: as certain areas come closer to achieving inviolable status, and designations of all kinds are taken more seriously, we might expect the process of site selection to be further politicised, and conflict and trade-offs to be shifted to that arena. Such a shift has indeed been apparent in some cases, with variable outcomes. An SSSI in Cardiff Bay was excluded, for example, from the proposed Lower Severn Estuary SPA (Pullen 1994) at least in part because urban redevelopment might be jeopardised by the stringent controls that such status would impose. More generally, there can be little doubt that the powerful provisions of the Habitats Directive have been a factor in delaying the nomination of SACs in various parts of Europe: four years after the deadline for submission of lists of candidate SACs (June 1995), thirteen member states had still not sent all the necessary information (WWF undated). Meanwhile, infrastructural and agricultural projects (often with EU funding) continued to damage important habitats; hence the accusation by some environmental groups that Europe had a fine system on paper while little was changing in practice (WWF undated, 1999). But there are countervailing examples. In an important ruling in 1996, the European Court of Justice held that the UK government had acted illegally in omitting an area of mudflats known as Lappel Bank from the proposed Medway Estuary SPA in Kent on the grounds that its ornithological value was outweighed by economic opportunities for port expansion.[36] The court insisted that economic requirements did not qualify as a general interest superior to the ecological objectives of the Directive; and even if economic considerations could, in some circumstances, be construed as 'imperative reasons of overriding public interest', they must be excluded from the designation stage, although they might be cited later to justify encroachment (Miller 1999b). For Lappel Bank this was a paper triumph. The mudflats had already been destroyed to make way for industrial development, without the rigorous tests that would have been required for part of an SPA. But important precedents were set, and interestingly the court's rulings in these cases have a distinctly non-instrumental and non-utilitarian flavour: planners have a role in invoking a duty to protect habitats and wildlife unless very special considerations apply. For the RSPB (1996: 1), the Lappel Bank case emphasised this 'crucial role' for planners 'in ensuring that the decision-making process required by European law is strictly followed'.

A tension between the strength and spatial extent of designation is reflected at local level too. While local authorities have taken more proactive interest in biodiversity at the sub-national scale, the UK government advises them only to 'apply local designations to sites of substantive nature conservation value, and to take care to avoid unnecessary constraints on development' (DoE 1994a: para 18). This has not prevented the declaration of increasing numbers of local nature reserves, but it is interesting to note that between 1980 and 1995 their average size fell from about 43 to 27 hectares (UK MAB Committee 1988). Attempts to delineate 'critical local capital' in development plans have in some cases been successfully challenged, although in others they have been accepted as supplementary planning guidance (in which case the tests come at the stage of development control). Paradoxically, a legal obligation to pursue sustainable development may make organisations wary of adopting definitions that could constrain growth: possibly for this reason, Scottish Natural Heritage eschewed detailed consideration of the environmental capital framework (but see Crofts 1998). Only a rigorous examination of the designation process at different scales could reveal a clearer picture, but, broadly speaking, the stricter the proposed protection, and the stronger the prospects of economic development as a potentially conflicting land use, the more likely it is that conflict will emerge at the stage of designation, or at the consultation stage of development plans. Despite the difficulties, however, and although it may not be as rapid or complete as some groups would hope, designation continues at international, national and local level.

In reviewing high-profile cases, it is easy to overlook the fact that not all conflicts between nature conservation and development are of the 'all or nothing' variety. It is sometimes possible for development to proceed without damage to habitats or species – perhaps even enhancing biodiversity – provided that certain conditions are applied and (crucially) adhered to. An RSPB survey (Brooke 1996), covering 560 cases in the UK between 1990 and 1995, concluded that planning conditions were 'increasingly seen as a means of overcoming nature conservation objections to proposals' (*ibid.*: 4). However, only 12 per cent of cases involved conditions to protect nature conservation interests or avoid potential adverse impacts; most used conditions for mitigation, accepting that some impact would still occur.[37]

Compensation

A related, and increasingly prominent, issue is that of the scope for creating or replacing habitats to compensate for those destroyed by development. Such compensation is a theoretical requirement of maintaining (non-critical) environmental capital and a legal one when damage to European sites is unavoidable. It is far from being a purely technical issue, however: rather, the question of compensation focuses attention once more on what is 'critical' and on the reasons for valuing particular habitats in the first place. If, for example, natural or semi-natural systems are deemed to have intrinsic value because of the processes that have produced them, rather than merely instrumental value by virtue of the 'services' they provide, then by definition they are not replaceable. The notion of 'maintaining environmental capital' through habitat recreation then becomes contentious – a clear case in which conflict is underlain by profoundly different convictions. There are other difficulties too: regarding a particular site as compensatable risks severing systemic ecological links and reinforces the concept of protection as tied to discrete parcels of land. Indeed, the 'integrity' of a site or system has proved something of a weasel word, around which reductionist and holistic views of ecological value have come into collision. To take one high-profile example, the Cardiff Bay Development Corporation spent more than £9 million in creating a wetland nature reserve as part of a package of compensatory measures to offset the loss of the Taff/Ely Estuary SSSI, where an amenity barrage would destroy a bird feeding habitat. While promoters of the new reserve defended it as 'a major asset' (Neal 1996: 3) from which 'economic benefits' would flow (Welsh Office 1997: 10), environmentalists' scepticism about a 'glorified duckpond'[38] reflected their view that what had been traded off was an integral part of a wider ecological system.[39] On the other hand, ScottishPower's enhancement of 10 square kilometres of hunting habitat for a pair of golden eagles 'to ensure that the birds have a net increase in prey should they choose to avoid the [Beinn an Tuirc] wind farm' (Scottish Power 1999: 11) in mid-Kintyre, Scotland, seems to have been broadly welcomed by the RSPB.[40] It is worth noting that where compensation requires the management of land, or financial commitments (for example, to fund a nature reserve), attaching qualifications to the granting of planning permission provides a key mechanism. For such purposes, planning agreements – legal and administrative instruments that enable permission to be linked to compliance with additional measures – are seen as more suitable than the simple use of planning conditions (Brooke 1996).

Although it would be counterproductive to ignore the potential of mitigation and compensation, it nevertheless worries conservationists that planning conditions and agreements (not always open to scrutiny) can disguise important losses:[41] there may be little monitoring, for example, and claims about translocation or recreation of habitats may prove, too late, to have been optimistic. It is significant, therefore, that in 1999, the Secretary of State upheld the decision of a planning inspector that in the case of an SSSI, the likely success or otherwise of the site's translocation was not a material factor in determining a planning application: 'SSSIs should be retained in situ, and translocation is … a last resort when faced with the inevitable loss of the SSSI …' (see Pulteney 1999: 12).[42] Other aspects of this case (in which Devon County Council had refused planning permission for the tipping of ball clay waste onto Brocks Farm SSSI, a small area of unimproved neutral grassland) are worthy of note because they show implicit appeal to differing ethical frameworks. The developer maintained that the SSSI was less important than the nationally important ball clay mineral and that economic, employment and competitiveness considerations favoured the tipping. However, the inspector argued that the particular type of grassland represented by the site was 'a rapidly diminishing resource' (*ibid.*: 11), and, significantly, that the acknowledged national importance of ball clay did not correspond to a need to permit 'at all costs' the proposal to create additional tipping capacity. Furthermore, feasible alternatives existed, and the appellants were criticised for not investigating these and for not proving a need for tipping space on the SSSI.

Progress and prospects

Definitive conclusions from our overview of policy and practice seem inappropriate at a time of rapid change, and in any case they would require more systematic assessment of policy implementation over time. But we can discern the broad direction of change, assess the extent to which it represents improvement and offer some thoughts about the future of nature conservation and planning. Certainly, one could say that more holistic approaches to land use and nature conservation have been gaining ground both nationally and internationally. At the same time, concern to promote sustainable development has been an important factor in a growing affinity between the relevant policy communities, as both have sought to interpret and apply the concept and have struggled with its contradictions. Especially in its earlier, more environment-centred incarnation, sustainability helped to raise the profile of biodiversity: environmental capital promised an attractive new framework for protection and compensation, and the objective of 'no net loss' resonated with more systemic thinking about conservation. Once again we conclude that sustainable development cannot resolve all conflict, although it may usefully show difficulties to be more apparent than real when too narrow, ignorant or short-term a view is being taken. More significantly, however, formulating policies for the protection of biodiversity becomes part of the project of deciding what is 'sustainable' and focuses attention anew on fundamental differences of interests, beliefs and values that may have become hidden in habitual exchange.

Conflicts have usually been played out, in the UK as in many other countries, within a framework in which the promised benefits of development have been seen as more tangible than, and have frequently 'outweighed', the benefits of nature conservation, sometimes with trade-offs resurfacing under the guise of environmental compensation. All too often, within this broadly utilitarian approach, even 'trivial short term pleasures' (O. O'Neill 1997:

131) have indeed triumphed over fragile natural values. This partly explains the continuing emphasis, in parallel with the growth of more holistic thinking, on the designation of special sites, and resistance to it by economic interests, sometimes supported by politicians. Designation remains an important way of giving nature conservation materiality in the planning process, and it allows planning authorities (and conservationists) to be one step removed from accusations of subjectivity: their role becomes essentially one of implementation. Fears that new thinking might undermine the institutional weight of designation are not entirely ill founded.

During the 1990s, however, the sustainability agenda (and changes in ecological thinking) began to undermine conventional appeals to the material benefits of growth as 'outweighing' the case for conservation. For one thing, within the broad framework of sustainable development, biodiversity itself acquired a neo-material rationale and therefore greater tangibility in its own right. To borrow Hajer's (1995) term, the storyline of 'biodiversity as crucial resource' was skilfully deployed in promoting tighter legislation and wider consideration of nature conservation objectives. One could argue here that sustainable development became a powerful discourse in the context of nature conservation, challenging dominant interests, although to exercise its discursive power, it had to speak the language that those with economic and political control were likely to find persuasive. Even so, as we have shown, the stronger legislation actually being driven through (and to some extent its interpretation in practice) cannot be fully rationalised in terms of instrumental or utilitarian thinking; in places, it reflects and implements a more ecocentric ethic, or a strong sense that appreciating intrinsic value in nature is constitutive of good human lives and societies. When, for example, it is ruled that economic interests cannot normally be cited to justify damage to important sites, or that compensation is an inappropriate substitute for conservation *in situ*, the implication is that certain benefits do not 'count' if, in order to attain them, an obligation to protect the site would have to be breached. Thus the institutional arrangements for the most important sites do not just give nature conservation greater weight; they effectively remove the habitat from an arena of simple trade-off, creating a framework in which judgement must be exercised about what human imperatives could be important enough to overrule the case for protection. Such thinking forces careful deliberation and makes it less likely that trivial pleasures – or, indeed, substantial material benefits – will trump fragile environments. One could argue that the legislation, however it may be presented, gives form to an environmental ethic that tempers the instrumental, utilitarian bias of markets and many government policies.

Of course, as we showed in Chapter 3, few frameworks, even consciously ecocentric ones, furnish an automatic warrant for conservation. Development that damages the natural world may in some circumstances be not only desirable but an ethical requirement. However, it will nearly always be open to question whether particular forms of development in particular places are the only or even the most appropriate way of meeting such a requirement. Interestingly, the need for the Cardiff Bay barrage to stimulate regeneration was challenged by community groups, politicians and conservationists throughout the six years in which a government decision was being made; as the inspector in the Brooks Farm ball clay case also recognised, there are often alternative ways of delivering economic benefits. This returns us to one of our recurrent themes: that claims about economic benefits must be open to more rigorous scrutiny; they need to be argued and defended, as has more usually been required of claims for conservation. Far from being beyond the remit of the planning system, such deliberation could be promoted as one of its central functions.

Effective land-use planning is crucial to nature conservation in a number of other ways. It is an important system for instituting appropriate frameworks and rules; it plays a key role in site protection; it can help to operationalise more holistic concepts of conservation; and it might contribute to the cultivation of respect for non-human nature. At the interface between planning and nature conservation, there are signs of the characteristics of such a system, but there is undoubtedly a very long way to go. We have been cautiously optimistic about the prospects for site protection, although it is important to remember that we have concentrated on countries with relatively sophisticated land-use planning systems, and we should not lose sight of what is happening at a global scale. Nor do we underestimate the sustained effort required to consolidate even modest gains. The most protective policies are implemented slowly and struggle to keep pace with development pressures, in part because economic interests fight to limit the coverage of such policies even when they have lost the battle on content.

However, the real challenge, recognised but as yet hardly grasped, is that of 'no net loss', or enhancement, of biodiversity. This goal is widely accepted rhetorically but remains contentious in theory and will certainly lead to conflict in implementation. It is this that will limit what is acceptable in terms of economic development, more so than the stricter protection of designated sites (although that remains essential), since it requires that significant damage is avoided or adequately compensated for *wherever* development takes place. Protection and enhancement of biodiversity 'in the round' demands robust land-use policies and widespread diffusion of holistic thinking from its modest beginnings in the 1990s: applying ecological guidance for development to much wider areas might be one way forward (Bishop *et al.* 1995). Most controversially, perhaps, this wider goal demands attention be paid to less direct ways in which planning could influence patterns of production and consumption for goods and services such as minerals, energy or transport, since these, ultimately, have profound implications for habitats and wildlife. While the dominant paradigms of the industrial system still enjoy the 'monopoly of social purpose' noted by J.K. Galbraith in the last of his 1966 Reith Lectures (Gregory 1971: 199), this will not be easy. However, we suggest that while the lines of conflict remain clearly drawn, the changes reviewed in this chapter are of sufficient significance to represent a challenge to this monopoly of purpose in terms of the ways in which competing claims are resolved.

7

DISTRIBUTING DEVELOPMENT
Sustainability and equity in minerals planning

The planning system has a key role ... in ... helping to
provide for necessary development *in locations* which do not
compromise the ability of future generations to meet their
needs.

DoE 1997a: para 39, emphasis added

... there is no conceivable system of ideas about justice which
will resolve all conflicts in a perfectly fair manner and against
which no reasonable argument can be mounted.

Low and Gleeson 1998: 197

... but where to draw the line between [the particular and the
universal]? ... I am drawn to the centrality of process, to ask to
what degree can we mobilise tolerance, negotiation and respect
through inclusive processes to reach acceptably just ends?

Munton 1999: 25

Introduction

Planning for minerals raises concerns that are central to sustainable development. The extraction, transport and consumption of minerals are significant
causes of environmental change (Plant *et al.* 1998), creating pollution and
threatening valued landscapes and habitats. Consequently, mining and quarrying have long attracted environmental opposition (Agricola 1950;
MacEwen and MacEwen 1982). Set against these concerns, it is frequently
argued that minerals are integral to modern urban fabric, if not to civilisa-
tion itself.[1] The broad concept of sustainable development would appear to
be vital in resolving conflicts over minerals development, but working out
what is sustainable in this context has often proved vexing, illustrating
many of the issues identified in earlier chapters. While the long-term environmental effects of minerals working can be positive, since many valued
landscapes owe their existence to mining or quarrying by previous generations, a difficulty in the context of potential mining projects lies in working

out *a priori* whether particular environments could be meaningfully restored. As we argued in Chapter 6, this is substantially an issue of environmental values. In determining how far to accommodate minerals production or protect the environment, planning encounters divergent views about needs, demands and market freedoms, similar in many respects to the core debates of transport policy. In this chapter, we examine how far conceptions of sustainable development could underpin spatial solutions to these conflicts, in siting minerals development and in linking together different scales of decision making.

Reconciling the interests of multiple spatial scales and locations has regularly presented planning with some of its most intractable dilemmas. Sustainability adds another dimension to this problem without necessarily providing a coherent framework for resolution; indeed, its apparently global currency begs questions about the meaning of 'global' and about relations between priorities and responsibilities at different scales (Reed and Slaymaker 1993; Yearley 1996). Logically, sustainable development implies changing material practices in every locality, yet conflict persists over how much autonomy nations or smaller communities should enjoy in defining what is sustainable within their territory when 'local' decisions have (sometimes unforeseeable) consequences elsewhere (Munton 1999). Identification with the 'national interest' has traditionally enabled economic development (and in some cases environmental protection) to trump competing demands, but interpretations of sustainable development raise questions about social purpose as well as scale, and about obligations to distant others. In effect, operationalising sustainable development entails a familiar – but formidable – task for government: that of reconciling general principles and particular circumstances. Simplistic prescriptions are unlikely to prove adequate to this challenge.

In this chapter, we examine three sets of issues that are readily apparent in minerals planning but each of which has broader relevance for our argument. The first concerns connections between sustainability and processes of uneven development. We suggest that approaches to sustainable development seeking to influence social and spatial differences can, equally, be profoundly shaped by them. In the context of minerals, this occurs because questions about *where* and *how* extraction should take place are linked to questions of *whether* certain resources should be exploited; similar links have been made in debates about transport (see Chapter 5), housing (Bramley and Watkins 1995), hazardous waste (Gerrard 1994), water (Rees and Williams 1993), and energy (Thayer 1994).

The second set of issues revolves around tensions between 'central' objectives and 'local' agendas, and the extent to which promoting sustainable development might help to resolve them. This is closely related to our third set of issues, concerning environmental justice. Clearly, selecting one site for minerals extraction over another has implications for *intra*-generational justice in the distribution of economic and environmental goods and bads, for many commentators a key dimension of sustainable development. The resulting impacts on environmental quality, and on the economic prospects of extractive communities, also have implications for *inter*-generational justice. Minerals planning shows clearly how intra- and inter-generational dimensions of justice can become closely connected, although they cannot automatically be rendered compatible. The question, then, is how far planning for sustainability can assist in aligning these different dimensions of justice by articulating the economic and environmental priorities of diverse localities across different levels of government.

We develop these arguments by examining the British minerals planning system, especially planning for aggregates – crushed rock, sand and gravel used in bulk for construction purposes – and its treatment of sustainability. In some respects, this sector raises fewer issues of noxious long-range environmental pollution than either energy minerals or metals, yet the convergence of a number of trends – commitments to sustainability, the emergence of international trade in aggregates, and political devolution within the UK – have coalesced to present profound challenges to the traditional national and hierarchical structures of minerals planning.[2] To set the scene for these policy developments, we begin by sketching some more general links between minerals production, sustainability and uneven development.

Sustainability, uneven development and non-renewable resources

The relationship between uneven development and sustainability depends on how the political economy of environmental problems is understood and on how sustainability is conceptualised. Brundtland saw the causes of and solutions to social injustice and environmental degradation as intimately linked: 'inequality is the planet's main "environmental problem"' (WCED 1987: 6). However, to view further economic growth as the solution to inequality, as Brundtland did, may be misguided if (as Marxists and many political ecologists would argue) there is an innate tendency in capitalist modes of production to expand and chase profit-making opportunities from region to region, creating problems of social injustice, environmental degradation and resource depletion (Eckersley 1992; Faber 1998; Harvey 1982).

The politics of environmental protection is bound up with uneven development in complex ways. One effect is to stimulate 'place-based political coalitions that organise to enhance and protect local resource and production complexes, to appropriate gains and displace losses' (Roberts and Emel 1992: 249). Blowers and Leroy (1994) observe how the reciprocal processes of 'peripheralisation' can appear to reinforce inequalities. Less desirable developments tend to become concentrated in 'peripheral' regions, characterised as isolated from metropolitan centres, economically marginal (often with a high degree of dependence on single, externally controlled industries) and having a political culture of acquiescence towards polluting operations by major local employers. At the same time, politically powerful communities are able to displace locally unwanted land uses and sustain higher quality environments as a result. However, while correlations between the distribution of environmental degradation and social inequality have been widely acknowledged (Faber 1998), the links between environmental protest, siting issues and uneven development can rarely be understood in straightforward locational terms. A well-developed capacity for resistance to unwanted land uses in core locations cannot always be dismissed as NIMBYism[3] (Freudenberg and Steinsapir 1991): opposition increasingly questions the very existence of certain forms of development because of the wider threats that they pose. Such challenges link local disputes to debates about the overall direction and level of economic growth and have, in some contexts, been constructed in the languages of sustainability and environmental justice (Beynon *et al.* 2000; Faber 1998).

The significance of these political processes for planning is that assumptions about uneven development and sustainability are bound into debates about the legitimate role of different policy approaches. Traditionally, planning has involved spatial and technical 'fixes', which seek to reconcile competing demands by influencing the location of development and managing environmental impacts. Alternatively, governments and others can pursue 'structural fixes', which challenge established patterns of production and consumption, if necessary by tempering demand. Concern for sustainable development has brought all these approaches to the table, but the balance between them is contested, with interesting spatial consequences. Where policies seek to maintain prevailing patterns of activity and accommodate trends, this requires that at least some regions receive industrial, extractive or waste-dumping activities (Lowe *et al.* 1993). If amenable locations can be found – through coercion, compensation, persuasion, local acquiescence or positive support – and defended as sustainable, then governments can defer difficult, systemic questions about growth. Alternatively, pursuing structural fixes – managing the *demand* for certain products – might ease the locational dilemma but raises awkward questions about projected 'needs' for particular goods and services.[4] Clearly, the scope for distributing development over *space* is bound up, in the policy process, with the sustainability of development trajectories over *time*.

In policies for non-renewable resources such as minerals, interactions between spatial, technical and structural fixes can be identified. It is inevitable, as a matter of definition, that such resources can be depleted, and one traditional policy response to scarcity has been to encourage and support the exploitation of further reserves (J. Rees 1990). Another response, given renewed emphasis by concern for sustainable development, has been to support structural fixes to moderate demand, focusing on the efficiency of resource use, recycling and the timely development of substitutes (Jacobs 1991). For aggregates these perspectives come together, since the pressures on governments to consider new spatial or structural fixes stem less from concerns about overall geological depletion than from the need to regulate the impacts of production on valued environments (UK Government 1994a). Much depends, therefore, on whether such environments are treated as strategic constraints on development or as considerations to be weighed against the demand for minerals; much also depends on the power of the arguments for different approaches to reconciling economic and environmental needs. Depending on how effectively these can be promoted by actors at different levels of government, outcomes may tend towards the displacement of extraction to more 'acceptable' locations, towards ever more careful management of extraction sites, or towards managing the demand for primary aggregates materials.

Much of this debate has taken place within the planning system, which provides an important institutional framework linking site-specific regulation and broader structural change. Planning law and policy influence the distribution of economic opportunities between activities and places, and allocate decision-making power between localities, interests and tiers of government. As we noted in Chapter 4, although planning has become an arena in which meanings of sustainability are contested, the choice of solutions is not wholly open: it is still largely framed within a traditional land-use remit. Within this framework, however, the planning system structures the relationship between minerals and the environment, sometimes treating the latter as a constraint on resource extraction, in other cases seeing environmental conditions as negotiable. Because planning permissions (including mining consents) are rooted in property rights, which can be costly to renegotiate as social priorities change, these too can exert inertia in the transition between past and future environmental conditions. The overall result is a degree of ambivalence towards patterns of uneven development. Planning policies tend to reinforce the concentration of industrial activity by treating land already degraded as a suitable location for further potentially damaging development, while 'more natural' landscapes may be codified as critical environmental capital meriting tighter protection.

In the context of uneven development, it seems inevitable that some conceptions of sustainability will resonate with local political cultures more than others. We might predict that environment-led interpretations, mobilised through planning, will shift undesirable developments towards communities whose economic difficulties predispose them to trade off jobs against environmental quality. However, 'local' decisions about minerals extraction must be reconciled with regional, national and international concerns, so that conceptions of sustainable development become entangled with the priorities of different tiers of government. Consequently, arrangements to link the different levels of planning have important implications for justice, both substantive and procedural. All of these arguments are well illustrated by the UK aggregates planning system, to which we now turn.

Minerals planning in the UK: supply and sustainability

The political importance of the aggregates and construction sectors is reflected in their sponsorship by the Department of Environment, Transport and the Regions (DETR) and the evolution of a detailed regulatory system for construction minerals. This system is operated primarily by local minerals planning authorities, which are responsible for drawing up minerals plans and determining individual applications for extraction.[5] Authorities have been expected to control the environmental impacts of quarrying by influencing the location of extraction sites and, through planning conditions, the timing and methods of production. Local planning decisions have also been guided by a detailed set of national policies and technical guidance, which emerged in conjunction with changing patterns of aggregates supply and growing environmental opposition to aggregates working. As local planning decisions increasingly restricted the development of new extraction sites, especially in south-east England, so both the industry and government began to express fears about future scarcity.

The policy response was to introduce measures to achieve 'an adequate and steady supply of materials to meet the needs of the construction industry at minimum money and social costs' (DoE 1976: 1). This core policy principle has been effected through the development of a broadly hierarchical planning framework. At the apex are national aggregates demand projections, based on models of future construction investment (linked to GDP) and its relationship to primary aggregates consumption.[6] Regional apportionment of projected demand is then considered by specialist bodies in each region, Regional Aggregates Working Parties (RAWPs), in consultation with a National Co-ordinating Group (NCG)[7] in the light of regional demand, inter-regional movements and deposits for which there is already planning consent. The RAWPs are charged with producing technical analyses for the forthcoming plan period, which are then used by central government to inform national and regional guidelines for aggregates provision (DoE 1989; 1994d). Each RAWP also assists the local minerals planning authority in dividing the regional guideline figures between them, and these sub-regional apportionments are used by minerals planning authorities in preparing their minerals local plans.

Through this hierarchy, the government has sought to co-ordinate potentially divergent local aspirations around the 'dominant strategic line' (Jessop 1997b: 13) of sustaining low-cost and abundant supplies of aggregates. By the late 1980s, the system was operating as a form of 'predict and provide', in which demand calculations and apportionments were treated as the kind of 'black box' that we identified in Chapter 4. The complex technical basis of the projections meant that they were difficult for individual planning authorities or environmental groups to question, and the largely closed, 'apolitical' culture of the RAWPs made them generally ill disposed to opening up the underlying assumptions (Cowell and Murdoch 1999). Minerals planning policy guidance reinforced the status of the apportionments and projections, advising that minerals planning authorities should 'make every effort to make provision for the supply' (DoE 1989: para A2). The overall effect was to entrench assumptions about the level and resource intensity of economic growth at the heart of minerals planning policy. However, while the planning framework has evolved to ensure that local land allocations support regional and, in turn, national aggregates requirements, it has also served to accommodate different regional circumstances through spatial fixes. The declining self-sufficiency of south-east England, for example, has been accompanied by flows of marine-dredged aggregates and crushed rock from large quarries in adjacent regions.

Environmental concerns were not neglected during the postwar period; nor were they wholly the preserve of 'local' planning. Alongside the strategic supply framework, the government produced a veritable encyclopaedia of Minerals Planning Guidance notes (MPGs), setting out the criteria for extraction affecting designated landscapes, ecologically sensitive areas and other important environmental assets, such as high-quality agricultural land. It also extended the capacity of the minerals planning authorities to regulate environmental effects, notably by increasing their powers to secure land restoration through planning conditions. But these advances in spatial and technical regulation did little to stem environmental objections to minerals production. National parks continued to yield large volumes of aggregates, transport remained a source of conflict, and the legacy of old minerals permissions (some granted during wartime with minimal consideration of environmental implications) diluted genuine improvements in the environmental performance of new sites. By the early 1990s, the structure of the minerals planning system itself had come under attack, with environmental groups criticising the way in which rising projections drove the allocation of ever more land for aggregates extraction. In fact, in the late 1980s the projections were underestimating the booming demands of construction activity, prompting the minerals industry to raise again the spectre of impending resource shortfalls.

These conflicts came to a head in the early 1990s when the government reviewed its strategic policies for aggregates provision in England – Minerals Planning Guidance Note 6 (MPG 6).[8] So intense was the controversy that the government decided to defer many of the key institutional questions until after MPG 6 had been revised, making a commitment to review the overall approach to minerals planning. Nevertheless, the government felt able to announce that the revised MPG 6 (published in 1994) heralded 'a gradual change from the present supply approach' in order to meet objectives 'in a way which is consistent with sustainable development' (DoE 1994d: para 25). A similar shift in philosophy has since been stated prominently in other revised Minerals Planning Guidance notes. However, it is clear from the guiding principles of minerals planning, set out in Table 7.1, that the government found it very difficult to respond simultaneously to pressures from the industry to sustain 'an adequate and steady supply' of aggregates and pressures from environmental organisations and some minerals planning authorities to prioritise environmental concerns. Certainly, environmental groups had won some battles over the language of policy: structural fixes were given greater emphasis than hitherto, and the government now clearly supported the careful use of primary resources, the recycling of waste aggregates and the greater use of 'secondary' materials such as mining spoil and china clay sands. Yet little consideration was given to demand management in the sense of reducing the *consumption* of aggregates. An important policy core remained intact, notably in the fundamental belief that 'it is essential, in order to contribute to the long run performance of the economy, that there is an adequate and steady supply of minerals' (DoE 1996b: para 3). However, rationalising this objective with a defensible interpretation of sustainable development had become more difficult, given that the projections used to define 'adequate and steady' supply anticipated a near doubling of primary aggregates consumption in England and Wales, from the 220 million tonnes consumed in 1991 to between 370 and 440 million tonnes in 2011. Environmental groups have claimed that seeking to identify sufficient sites to accommodate this level of demand 'forced vast

Table 7.1 The principles of sustainability in government minerals planning policy

---

(i)   to conserve minerals as far as possible, whilst ensuring an adequate supply to meet needs;

(ii) to ensure that the environmental impacts caused by mineral operations and the transport of minerals are kept, as far as possible, to an acceptable minimum;

(iii) to minimise production of waste and to encourage the efficient use of materials, including appropriate use of high quality materials, and recycling of wastes;

(iv) to encourage sensitive working, restoration and aftercare practices so as to preserve or enhance the overall quality of the environment;

(v) to protect areas of designated landscape or nature conservation value from development, other than in exceptional circumstances and where it has been demonstrated that development is in the public interest ...; and

(vi) to prevent the unnecessary sterilisation of mineral resources.

Source: DoE 1996b: para 35

swathes of countryside to be earmarked for destruction by unnecessary quar-rying' (CPRE 2000b: 2) and avoided any strategic recognition of environmental constraints on supply (Plowden 1994).

Central government leaned heavily on spatial and technical fixes to give coherence to its interpretation of sustainable development, with significant distributive consequences. Further encouragement was given to improving standards of quarry design, management and reclamation. Indeed, the government and the minerals industry have at times been almost evangelical about the scope for careful quarry restoration to leave an environmental asset 'of equal or added value' to future generations (UK Government 1994a: para 18.25). Presumptions against extraction in designated areas were reinforced, and planning guidance expressed a carefully qualified 'preference for extending existing workings ... as a means of minimising environmental disturbance' (DoE 1996b: para 61). At a broader scale, the government sought to use the planning system to reduce the proportion of supply from 'traditional land-won sources' in England from 83 per cent in 1992 to 68 per cent by 2006 (DoE 1994d: para 26), guiding local authorities with a set of revised regional apportionments. An increasing proportion of supply was also expected to come from 'imports from outside England and Wales'.[9] While somewhat reticent on the precise sources and timescales, the govern-ment (*ibid.*: para 44) believed that 'subject to tests of environmental acceptability an increasing level of supply can be obtained from coastal superquarries', of which more below. The revised MPG 6 also seemed to grant minerals planning authorities greater flexibility to adopt conceptions of sustainable development that departed from the dominant strategic line of national policy. Demand projections were regarded as a 'starting point' (UK Government 1994a: 124) for considering policy options, with 'the preparation of development plans [providing] an important opportunity to *test* the practicality and environmental acceptability *at the local level* of the Guideline figure' (DoE 1994d: para 58, emphasis added). However, our analysis in Chapter 4 indicates the naivety of expecting a process of local 'testing', of itself, successfully to challenge core policy presumptions. In the next section we examine how central government has responded to local challenges to its minerals policies, but before turning to this, we review a number of changes to the national framework that followed publication of the 1994 MPG 6.

Reflecting on the system it inherited, the Labour government elected in 1997 considered that the 'broad approach' to sustainability in aggregates planning was 'fundamentally sound' (DETR 2000f: para 2.26). Nevertheless, structural change was beginning to achieve greater parity with spatial and technical fixes, as reflected, for example, in the view of Planning Minister Richard Caborn that 'the closer we can get to the concept of "aggregate supply" without distinguishing between newly dug, recycled or secondary material, the better' (ENDS 1998b: 2). The resource intensity of construction activity has also been seen as a legitimate target for policy intervention, although objectives have generally been pursued through instruments external to the land-use planning system.[10] These goals have been supported by a landfill tax, introduced in 1996, one effect of which has been to raise the cost of discarding unwanted construction material.[11] By the late 1990s, it looked as if the MPG 6 targets for utilising non-primary aggregates would be more than met (ENDS 1998c). Symbolically the most significant, perhaps, a government 'best practice guide' on sustainability appraisal suggested that regional planning bodies adopt targets not only for making greater use of recycled and secondary materials but also for the *reduced* consumption of primary aggregates (DETR 2000c).

However, responses to the government's proposals for a tax on virgin aggregates extraction showed that tensions between spatial, technical and structural fixes still ran deep, with arguments about sustainable development drawing upon quite different moral and political precepts. The proposed tax was justified as a means of internalising those environmental costs of extraction not mitigated by planning regulation, and also as creating a price incentive to favour secondary and recycled materials, which would be exempt (see Her Majesty's Treasury 1997). The minerals industry was implacably opposed, but the government declared that the tax would be introduced unless the industry could deliver a package of voluntary measures that provided at least as great an environmental benefit. This political pressure forced some movement on previously intractable issues, but the 'new deal' offered by the Quarry Products Association (QPA)[12] displayed a strong preference for spatial and technical measures. Commitments to tighter environmental management featured prominently and, in the context of hardening government attitudes towards extraction in the national parks, the QPA included measures that could reduce the development of resources in these areas. Some major companies additionally volunteered to relinquish old minerals permissions (Chambers 1999). But from the industry's perspective aggregates remained 'fundamental to our way of life' (ENDS 1998b: 2), such that managing demand would lead inevitably 'to control on the personal choices and freedoms which we all enjoy' (Harrison 1997: 10): sustainability was seen as a spatial and technical problem, entirely a matter of how materials are worked and where. Unconvinced, the Chancellor of the Exchequer announced (in the March 2000 Budget Statement) that the tax would be introduced in April 2002.

Overall, the 1990s was a period of significant uncertainty for the minerals sector and for the 'predict and provide' approach. While the projections indicated sustained growth, construction demand in the UK actually fell sharply, and aggregates production in 1998 was about 220 million tonnes, 80 million tonnes less than in 1989. Having completed the review of the overall approach to minerals planning promised by the government in the 1994 MPG 6, the consultants concluded that fundamental tensions remained between the emphasis on maintaining an adequate and steady supply and alternative approaches that sought to contain aggregates provision within environmentally sustainable limits (DETR 1998j). What makes this inconsistency especially significant is that even within the highly centralised arrangements of British minerals planning, the planning system is far from a passive transmitter of central government policy (Allmendinger and Thomas 1998). Although MPG 6 had managed to achieve a high degree of internal numerical coherence, its prescriptions were greeted with an array of supportive, critical and overtly challenging responses in the local planning system. To varying degrees, these local responses have in turn influenced the development of national policy. To illustrate this relationship, we now explore initiatives in environment-led minerals planning in south-east England, conflict over a proposed superquarry at Lingerbay in the Western Isles of Scotland, and the planning processes implicated in the international trade in aggregates.

Place, power and spatial scale

South-east England as a regional NIMBY?

Articulate opposition to unwanted development in the countryside across south-east England has been a constant in postwar planning. A combination of affluence, counter-urban migration, intense pressures for development and the cultural resonance of rural landscapes has prompted action to protect environments from minerals production. Arguably, south-east England has enjoyed greater success than the rest of Britain in translating environmental concerns into the spatial framework of national minerals policy, with relative reductions in regional self-sufficiency matched by aggregates imports from elsewhere (Lowe *et al.* 1993).[13] Given the context, it is not surprising that environment-led approaches to sustainability have proved especially appealing in this region (Counsell 1998), and there may be a suspicion that sustainable development provides powerful social groups with a positive ethical gloss for NIMBYism, legitimising the displacement of unwanted development to more accommodating locations. Nevertheless, the region continues to be a major producer and consumer of aggregates. Moreover, local minerals planning authorities have had their interpretations of sustainable development curtailed in line with national minerals planning objectives.

An early example is provided by Berkshire County Council's attempt to use environment-led concepts of sustainability in revising its minerals plan. The final selection of only certain 'preferred areas' in which there would be a general presumption in favour of permitting extraction reflected the planning authority's goal of maintaining the 'critical and compensatable environmental capital of the County' (Babtie Group Ltd 1993: para 123).

On this basis, it was concluded that allowing for aggregates production at the rates indicated in regional guidance after 1996 – 2.5 million tonnes per year – would be 'unsustainable' (*ibid.*: para 132) because it would breach environmental capacity. Sensitive to accusations that they were 'exporting unsustainability' (by displacing the effects of local minerals demand to areas outside Berkshire), the council incorporated policies in the draft minerals plan to reduce local demand for land-won aggregates by minimising waste, making the most appropriate use of high-quality materials and increasing the use of recycled and secondary aggregates. At the ensuing public inquiry, however, the minerals industry argued that local minerals planning authorities had neither the capability nor the remit significantly to influence patterns of aggregates consumption, and that policies for promoting recycled and secondary materials were over-optimistic – criticisms that the inspector accepted. Both the industry and neighbouring local authorities questioned the fairness of Berkshire's unilateral decision not to meet its apportionment, expressing fears that adjoining counties would have to make up for any deficit. The inspector concluded that if other authorities adopted Berkshire's interpretation of sustainable development, 'severe constraints would be placed on the production of aggregates which have a vital role to play in the national economy' (Brundell 1994: 13). Such arguments reinforced a straitjacket of vertical relations between national government and local authorities and effectively restricted local conceptions of sustainability to the accommodation of apportionments by siting and technical means.

Since the Berkshire inquiry, the government has come to endorse the 'testing' of national guidance in development planning, but there is little evidence that local deviations from quantified aggregates requirements have become more easily defensible as a result. West Sussex County Council's environmental capacity study (see Chapter 4), for example, 'led to some fundamental changes in the strategy for mineral working' (West Sussex County Council 1997: para 2.22). As with Berkshire, the council concluded that it would be unable to meet its minerals apportionment in full without causing unacceptable environmental damage. When the draft minerals local plan went to inquiry in 1998, the inspector concurred with the council that local minerals planning provided an opportunity to 'test' the apportionments and did not disagree with the broad approach. However, he was less convinced that the constraints leading to the omission of certain sites were 'insurmountable'. Moreover, because guidance requires that plans make 'an appropriate contribution to meeting local, regional and national needs', there was a requirement for additional supplies from the county (Mead 1999: para 5.317). He recommended the inclusion of further extraction sites in the plan to meet the levels set out in national and regional guidance.

These examples suggest that it is difficult to base local plans on conceptions of sustainable development that require deviation from the priorities of national policy. Councils wanting to depart from their allocated minerals apportionment must confront the technical basis of demand projections, the integrity of jointly achieved regional decisions and the political importance accorded to accommodating aggregates demand in supporting economic growth. Although national policies conferring protection on certain habitats and landscapes can be translated into local minerals planning policies that influence the *location* of extraction,[14] spatial constraints cannot be extended to redefining the level or material content of a sustainable aggregates supply. However, it would be simplistic to conclude that actors within local minerals planning arenas are impotent; what is interesting is that they tend to enjoy greater success when they articulate their concerns in higher-level political arenas. Changes to minerals apportionments and national policy do reflect political pressure from individual authorities and objectors (Murdoch and Marsden 1995): local interpretations of sustainable aggregates production can be translated into planning policy when agreed regionally, passed 'up' to be legitimised by central government and passed back down the planning hierarchy. In effect, this is what happened with Berkshire's minerals plan: pressure from minerals planning authorities across the south-east during the MPG 6 review prompted central government to endorse lower regional apportionments. These translated into a *pro rata* reduction in Berkshire's allocation for the period 1992–2006, thus delivering the county's original objectives, but by a different route.

None of this is to overlook the genuine problems that might arise if local planning decisions could thwart the supply of materials essential to meeting important social needs. However, until the late 1990s, the government (like the industry) tended to equate all projected demand with a need that ought to be met. Consequently, it pursued a spatial fix that linked supplies from 'alternative' locations and sources to its overall supply objectives.[15] The extent to which exploiting spatial variations in the environmental, geological and economic acceptability of aggregates extraction could be regarded as sustainable has been a key element of debates over coastal superquarries, our next – and closely connected – example.

Remote coastal superquarries: the best environmental location?

Coastal superquarries can be defined as extremely large crushed rock quarries, with reserves of at least 150 million tonnes and an annual output upwards of 5 million tonnes, which is mostly exported by sea (SOEnD 1994). The concept seems to exemplify powerful trends in the uneven development of mineral resources, where continued dependence on current modes of production – seeking to meet rather than manage demand – means postponing crises of supply 'through the spatial expansion of extractive industry into increasingly marginal areas' (Emel and Bridge 1995: 324). Coastal superquarries first attracted serious attention in the UK during the 1970s, as the government began investigating ways of addressing potential supply shortfalls in south-east England. Only one company, Foster Yeoman, made any significant investment, developing the Glensanda superquarry in the Western Highlands of Scotland. However, as the government found it necessary to respond to mounting environmental pressures over 'traditional land-won sources', so coastal superquarries became woven into the strategic line of national minerals policy. Research published by the DoE in 1992 indicated an economic and geotechnical potential for superquarries around the coasts of Scotland and Norway to supply markets in south-east England (Whitbread and Marsay 1992). The DoE examined the proportion of English demand that might feasibly be met from 'imports' between 1992 and 2006 and included an indicative figure of 160 million tonnes in MPG 6. At the same time, the Scottish Office wrestled with producing planning guidance for site selection in Scotland, which provided support in principle for up to four coastal superquarries and specified the 'exceptional circumstances' for permitting quarries in nationally designated areas (SOEnD 1994: paras 18 and 19).

Coastal superquarries could be seen as a spatial fix that has helped to support the numerical integrity of a demand-led minerals planning strategy, but it has also been rationalised in terms of sustainability (see DoE 1976; SOEnD 1994). The argument has been that concentrating extraction at a few high-output sites reduces the unit environmental impact (in terms of transport, dust, noise and vibration per tonne of aggregates produced) compared with an equivalent output delivered from many smaller, inland quarries; that sea transport is environmentally preferable to moving aggregates by road; and that fewer people would be directly affected by quarrying in remote, rural locations. These arguments make a number of assumptions about environmental value and uneven development, with significant effects. Correlating 'impact' with the number of people having direct sensory experience channels extraction to lightly populated areas and downplays the value of relatively 'wild', unindustrialised landscapes.[16] Consequently, the government's efforts to minimise the overall environmental impacts of meeting aggregates demand, put forward as the 'best balance' in the *national* interest, legitimises the redistribution of environmental impacts to potential superquarry locations.[17] The further contention that job-creating investment of this nature ought to prove more acceptable for fragile economies that need economic development (as discussed by Mackenzie 1998) highlights the interactions between processes of peripheralisation and the balance struck in the name of sustainable development.

Unsurprisingly, therefore, the coastal superquarry scenario has attracted considerable criticism. The DoE's initial preference for meeting a much higher proportion of English demand from 'imports' was moderated following lobbying by environmental groups as well as by sections of the aggregates industry. Most fundamentally, perhaps, the quest for raw materials in peripheral locations is likely to encounter social groups with distinctive value systems and priorities that cannot easily be reconciled to 'national' resource agendas (Emel and Bridge 1995). Across Scotland, for example, nationalist and environmentalist objections intertwined, identifying policies supporting coastal superquarries as 'a breathtaking piece of Anglo-centrism' (Dunion 1992: 54) that displaced English unsustainability into Scotland's backyard. While geographically peripheral communities often have little political or economic power, they can sometimes draw upon the fact that the environments they inhabit have acquired wider ecological and cultural significance. Many sparsely populated and mountainous landscapes around Europe are valued for their relatively undeveloped qualities and special habitats, values increasingly institutionalised in national and international conservation policy. This confounds the traditional 'national need' versus 'local impacts' dichotomy of planning conflicts and again makes determining what is sustainable a problem for all levels of government. Nor is the issue straightforwardly one of environmental costs versus economic benefits. Scottish guidance advised planning authorities to consider not just the jobs created by superquarries but also potential damage to tourism and fishing, and the risks of 'disrupted lifestyles and social cohesion' (SOEnD 1994: para 61).

In practice, the power of national minerals planning policies depends on whether they can be translated effectively into site-specific planning decisions. The conflicts over the Lingerbay superquarry indicate some of the difficulties. In March 1991, Redland Aggregates Ltd[18] applied for planning permission to extract approximately 550 million tonnes of anorthosite rock from the mountain of Rhoineabhal at Lingerbay on the island of Harris in the Western Isles. Extraction would continue for a sixty-year period and would supply general-purpose aggregate to markets in southern England and continental Europe. Public opinion on Harris has been and remains divided (Mackenzie 1998), but the application also attracted objections from Scottish governmental and non-governmental environmental organisations, as well as from individuals across the Western Isles, Scotland and beyond. As the arguments rumbled on, the Secretary of State for Scotland called in the application for what ultimately became Scotland's longest-running planning inquiry. The Lingerbay superquarry generated a complex and wide-ranging debate.[19] We focus here on the conceptions of sustainable development promulgated, through the planning process, by the various parties, and on the importance of spatial scale in giving weight to these different stances.

Arguments about jobs featured prominently in the public inquiry. Redland presented sustainable development as a national project of progress and growth (Mackenzie 1998), where providing jobs for a community with high unemployment and limited economic opportunities could outweigh the minimal environmental effects that a well-managed quarry would cause.[20] A need for economic development was recognised by the local authority, Western Isles Council,[21] which, initially at least, saw the quarry as a means of generating employment, checking depopulation and reversing the decline of the island's distinctive culture. A number of objectors became unhappy with the trade-off being put forward and disputed the scale at which costs and benefits should be compared. Cross-examination of the developer's economic witnesses revealed that while the Western Isles as a whole might gain 100 direct jobs when the quarry reached maximum output, far fewer would accrue to Harris, where environmental impacts would be concentrated. An abstract discourse of maintaining employment, income and population numbers into the future could not overcome concerns that the specific cultural and economic needs of existing islanders and their children might be threatened. Distinctive local religious and cultural practices, especially Sabbath observance and Gaelic speaking, were seen as endangered by the social changes that a superquarry would bring.

The significance of employment lies also in the hierarchical principles used to steer between conflicting policy objectives. 'Social or economic benefits of *national* importance' (Firth 2000: para 22, emphasis added) could provide a defensible planning reason, recognised in government policy, for permitting developments that compromised the 'objectives' and 'overall integrity' of designated areas of *national* importance: Lingerbay was located in a National Scenic Area (NSA) designated under Scottish legislation for its landscape qualities. Redland called expert witnesses to argue that a combination of careful siting, design and site management would actually leave the overall integrity of the area largely unaffected. For some objectors, however, landscape integrity was seen in different and distinctly material terms, which positioned the superquarry project as a fork in the path to sustainable development for Harris. Industries like Harris tweed, tourism, fishing, aquaculture and high-value pharmaceuticals were upheld as potentially more sustainable than quarrying (Link Quarry Group 1996), and representatives of each sector claimed that it was economically dependent for marketing purposes on an image of the Western Isles as 'pure, unspoilt, [and] environmentally clean' (Western Isles Labour Party 1994: 11). The proposed superquarry raised the spectre of peripheralisation (Blowers and Leroy 1994): should the perceived environmental degradation deter alternative sources of employment, people in Harris would find it more difficult in future to resist further damaging development rejected by other communities.

In their final analyses, both the Reporter and Scottish Ministers[22] supported Redland's claim that there was a national interest in securing economic development in remote rural areas (Pain 1999: para 27.69). However, Ministers ultimately gave greater weight to the adverse effects on landscape quality, arguing that such substantial industrial developments would inevitably alter the character of the NSA, no matter what technical fixes might be deployed. If the protagonists used different hierarchies of spatial scale to justify their arguments, the resolution owes much to the re-scaling of the planning hierarchy brought about by political devolution.

Redland had relied 'upon a statement of government policy that the encouragement of a specified alternative supply source, namely a coastal superquarry within the United Kingdom, is in the national interest' (Martin and Abercrombie 1995: 77). A central plank of this national interest was located in English minerals planning guidance: that the indicative figures for alternatives to traditional land-won sources in MPG 6 defined an anticipated shortfall that superquarries were 'needed' to meet, coupled with a stated policy preference for exploiting 'indigenous mineral resources' (SOEnD 1994: para 3). Such arguments reduced 'any pretensions Scotland might have to planning autonomy to a question of economic necessity in what is defined as "the United Kingdom"' (Mackenzie 1998: 514). They also provided another 'benefit of national importance' to be set against national (i.e. Scottish) site-protection policies. However, because of devolution, the decision on Lingerbay was not made by a Secretary of State for Scotland appointed by a government in London but by elected Ministers from the Scottish Parliament. Empowered to place Scottish interests at centre stage, Scottish Ministers could conclude that 'the need for stone in England should not override the need to protect Scottish areas of natural heritage value', and they rejected the application (Brown 2000: 5).

Aggregates trade and planning in a transnational context

With this dismissal, the likelihood of further superquarries in undeveloped, protected areas of Scottish coastline has receded.[23] The decision has wider implications for 'demand-led' resource planning strategies in England and, given growing interest in the international trade of aggregates, these implications reach beyond the borders of the UK. At the time of writing, imports meet only a fraction of British domestic consumption,[24] but the situation is very different in other parts of north-west Europe. Large volumes of aggregates are transported between the Netherlands, Germany, Belgium and France, and one benefit claimed for coastal superquarries is the capacity to ship crushed rock straight to urban markets in a range of countries. Assumptions about the potential geological availability and environmental acceptability of such supplies, and the political unacceptability of managing consumption, led numerous commentators in the early 1990s to anticipate international trade meeting a growing proportion of aggregates demand (Cowell 2000b).

Cross-border trade creates some challenges for traditional state-centred systems of environmental governance. Assertions of the 'national interest', used within states to mediate between competing objectives, have no warrant across national borders. Commitments to sustainable development did not lead national governments to investigate in detail the implications of their domestic planning decisions for environments in other countries. In the Netherlands, as in south-east England, public resistance to domestic sand and gravel extraction, articulated through the local planning system, has increasingly restricted domestic supplies. This pressure has led the Dutch government to seek to maximise the use of non-primary aggregates, including recycling and reducing construction waste. However, the overall consumption of construction materials is still expected to rise, encouraging officials to look positively towards international sourcing:

> Imported crushed rock from surrounding countries is a good, sustainable alternative for the Netherlands. Investigations have shown that the supply that can be expected from imports is suffi-cient for the long term.
>
> de Jong and de Mulder 1998: 213

What is 'sustainable' in this context seems to have more to do with what is economically and geologically available than with what might be environ-mentally acceptable or socially just. In so far as such policies have been based on cross-border communication, this has largely taken the form of discus-sions between state officials and an examination of planning policy documents in exporting countries. Another dimension of the Dutch political calculation is that securing imports would be more politically acceptable than coercing provincial planning authorities to produce more primary material (Ike 1999). Such calculations may not only prove inaccurate when quarry projects come forward in other countries but may also be challenged from an ethical perspective: the practice of seeking supplies from abroad to protect environments at home has been increasingly seen as 'exporting unsustainability'.[25] Responding to objections to the West Sussex Mineral Locals Plan, that sites in Scotland should be worked instead of local resources, the inspector explained that he had 'no evidence to suggest that the local environment in Scotland is any less precious than in West Sussex' Locals (Mead 1999: paras 3.4 and 5.30). Elements of peripheralisation can also be detected in the tendency for distant suppliers to be located in economically disadvantaged locations. An upsurge in exports from existing quarries in Poland and the Czech Republic to Germany reflected the lower aggregates prices, lower wages and weaker regulatory regimes that prevailed in the former Communist states (Cowell *et al.* 1998).

Crushed rock exports from Norway also increased steadily during the early 1990s, supplying southern England, Germany and the Netherlands, but these transfers are less easily characterised as exploitative. Although rural depopulation is a politically sensitive issue, few rural regions of Norway could be characterised as poor in absolute terms. Yet Norwegian development agencies have actively promoted the sparsely populated western coasts and fjords of Norway as a geological *and environmental* resource, arguing that they could be quarried at lower impact than in more densely populated parts of Europe (Norwegian Geological Survey 1997).

This 'eco-efficiency' rationale echoes the claim that Scottish superquarries are 'more sustainable', but there are crucial differences in the relations between national and local planning in the two countries. In Norway there is no strategic national planning framework for aggregates, and local government enjoys greater planning autonomy than is the case in Britain. Developers would find it difficult to use arguments of 'national economic need' to overcome local objections to minerals development.

This does not mean that quarry development is solely a matter of local community preferences. Quarrying projects can expose tensions between regional and national conservation policies on the one hand, and efforts to exploit the remoteness of the fjords and islands for international mining investment on the other. As actors at local, regional and national levels of Norwegian government debate whether economic or environmental needs should be met, distinctive patterns of export quarry development have emerged. Resistance to the development of coastal quarries has been most successful when new proposals are made for green field sites, where the effects on land, tourism and 'undeveloped nature' have provided powerful arguments against extraction. Computer maps showing the 'depletion' of large, contiguous areas 'untouched' by modern technology have attracted attention (see Berntsen 1994: 20–1), especially in the southern half of Norway, where few such areas are deemed to remain. At Jondal on Hardangerfjord, for example, although the municipal council supported coastal superquarry development, groups on the opposite side of the fjord and from the regional offices of state environmental administration were able to resist the project. Superquarry applications have proved more acceptable in municipalities already economically and culturally connected to minerals production. The Tarmac company secured significant local support for a superquarry in Sokndal municipality, where one-third of the employment was already in mining, and the site itself was relatively inconspicuous.[26] Elsewhere in Norway, much of the increase in crushed rock production for export has come from the expansion of smaller coastal workings or the incremental development of quarries from industrial sites, where local councils have viewed export quarrying development as a way of sustaining their income and employment base. In terms of our broader arguments about environmental justice, it is interesting that pre-existing industrial development generally persuades environmental groups that few assets of importance are at stake.

To conclude this section, it is clear that planning decisions struck in or for specific localities have implications for places and planning strategies elsewhere. In making sense of such interactions, however, the Norwegian experience suggests that accusations of 'exporting unsustainability' can fail to differentiate the exploitation of distant and vulnerable communities from the legitimate location of production in sites that are environmentally and socially acceptable. The issues are challenging: how best to reconcile claims about economic and environmental needs in different jurisdictions and at different levels of government; and how best to evaluate the merits of aggregates production in specific places against the scope for reducing demand. We now turn to an examination of how far conceptions of sustainability can align these different levels of interest, and to a consideration of the role that planning might play as a dialogical forum, along the lines that we suggested in Chapter 3, in negotiating multiple claims about sustainability, justice and political autonomy.

Scales of justice

Our analysis so far has provided some insights into broader patterns of uneven development in the aggregates sector and the interactions with earlier rounds of economic and environmental differentiation. When it comes to individual investment decisions, economic, political and environmental arguments tend to favour the concentration of minerals extraction in locations already intensively exploited, where communities are less willing or able to restructure their economic base. As these arguments unfold in the planning system, so patterns of inequality tend to be reproduced over time, and processes of peripheralisation can be detected.[27] Significantly, a major effect of sustainability discourses to date seems to have been to intensify such patterns of differentiation, encouraging the spatial concentration of minerals production while protecting critical environmental capital elsewhere from extractive activities. These effects should be viewed in part as a result of the traditionally limited capacity of the minerals planning system to move beyond spatial and technical fixes to engage in restructuring wider patterns of production and consumption. Nevertheless, we have also shown how sustainability discourses have been used to challenge the preferred spatial fix of governments, problematising claims that certain locations are objectively more sustainable for minerals production than others and, in turn, the overall acceptability of spatial and technical fixes as means of maintaining supply.

Finding the best location for necessary development clearly has a legitimate role in making development more sustainable, but determining what is 'best' and how far certain materials are 'needed' have become thorny political questions. Identifying apparent unsustainability and distributive injustices is relatively straightforward in comparison with working out more defensible policy strategies. Strategies for ecological modernisation in the minerals sector – reducing waste, controlling pollution, improving the materials efficiency of construction, and substituting recycled and secondary aggregates for primary materials – have an important role to play but will still leave large quantities of primary aggregates to be produced from somewhere. Moreover, ecological modernisation, like much theoretical environmental writing, has been concerned with total production 'rather than its disaggregated distribution' (Dobson 1998: 13) – a problematic omission given that, as we have shown, the level and location of minerals production are closely intertwined.[28]

One might hope that the concept of sustainable development could guide us, given that intra- and inter-generational justice are both regarded as integral to many conceptions. However, justice is subject to different interpretations, which may not automatically be compatible with particular conceptions of sustainability (Dobson 1998). Indeed, justice is the subject of such a voluminous literature – based on principles of rights, deserts, needs, equality and others – that we could never hope to do even minimal service to the relevant debates (see Dobson 1998; Low and Gleeson 1998; O. O'Neill 1996; Rawls 1972 for more detailed discussions). If we extend the 'community of justice' beyond the human world to encompass non-human entities, as we discussed in Chapters 3 and 6, then the relationships between justice and sustainability become even more complex.[29] In our more modest discussion, we interpret justice in terms of the fair spatial distribution of environmental goods and bads between human communities,[30] and in terms of procedural justice in the making of policies, plans and decisions. An important theme of our argument is that issues of space and scale are relevant to both distributive and procedural dimensions of justice; these dimensions come together in effecting arguably one of the core tasks in planning for sustainability, 'a just distribution, justly achieved'.[31] Our first step is to consider the guidance offered by environment-led conceptions of sustainable development.

Environmental capital and spatial scale

Justice is an important element of environment-led approaches to sustainability, but, as we argued in Chapter 3, they provide only sparse guidance on the important distributive questions that regularly emerge in planning for sustainability. Approaches based on maintaining environmental capital have been concerned mainly with *inter*-generational justice – ensuring that future generations inherit opportunities for well-being comparable with those of the present. There may be circumstances in which maintaining environmental capital would improve intra-generational equity, where that coincides with meeting the basic environmental needs of the poor (Pearce 1988). In supporting arguments for sustaining global life-support systems, it can also underpin needs that are a basic precondition for other dimensions of justice. But to assert that maintaining environmental capital necessarily promotes social justice begs questions about the dimensions of environmental value under consideration, and the ways in which institutions distribute the material and non-material benefits that flow from it. These questions arise in conflicts over minerals extraction in protected landscape areas, where it is often argued that, should the case for environmental protection prevail over that for development, the local community ought to receive some form of compensation for the loss of economic opportunities.[32] Clearly, mere proximity between capital (whether environmental or human-made) and some community does not necessarily make for a just distribution: consideration must be given to matters of entitlement.[33]

Applying capital- or capacity-based models also requires judgements to be made about the status of geographical differences in environmental quality. We have noted how, in practice, concentrating minerals extraction in areas already engaged in this activity can support the protection of critical environmental capital elsewhere, yet this seems unjust if it means that intensively quarried environments are treated as a low-cost opportunity for sustaining production. On the other hand, redistributing extractive facilities to 'pristine' and sparsely populated environments scarcely combines the requirements of inter- and intra-generational justice more successfully. One might respond to the reality of environmental and social differences between regions by arguing that just treatment is not synonymous with sameness of treatment, or equal outcomes; what justice requires is that we find 'morally relevant reasons' for treating regions differently (Shrader-Frechette 1985: 221–2). These reasons might take a number of forms, and one should expect them to be contested. In the context of minerals planning, quarrying interests might justify an uneven protection of environmental capital with the maxim that 'minerals can only be worked where they occur', but since this cannot amount to a moral injunction to work them, concepts of justice identifying a need for the material might better provide a morally relevant reason. Another approach could be to apply Rawlsian principles and seek to avoid locating environmentally damaging development where it would add to the burdens of the least well-off (Munton 1999).[34]

In a number of respects, therefore, assumptions about scale are integral to interpreting sustainability and intra-generational justice. These assumptions are bound up with particular conceptions of the community of justice, which affects to what or to whom justice is given (Dobson 1998; Walzer 1983). Economistic models of sustainability tend to view the spatial dimensions of this community in simplistic terms, assuming shared interests in global environmental sustainability. Yet many forms of development, including minerals production, generate impacts and benefits at multiple spatial scales – involving issues of residential amenity, landscape, water quality and wildlife protection – around which shared interests can be more difficult to discern. It is often assumed that being excluded from a particular community is problematic: thus we noted how aggregates planning in England and the Netherlands made assumptions about the environmental capacities of distant places to provide domestic markets with aggregates. In practice, however, inclusion within a political community does not eliminate the possibility of outcomes that might be considered unjust. Drawing up minerals planning strategies for the UK as a whole enabled the government to mount utilitarian arguments that favoured a few coastal superquarries in sparsely populated parts of Scotland in the overall 'public interest', over dozens of gravel pits in more densely populated parts of England (or, implicitly, over managing demand). Clearly, then, adjusting the scale of the political community has distributive implications in itself: by helping to define the scope for spatial, technical and structural fixes, it frames the 'decision space' within which economy and environment can be reconciled. These distributive implications, and the tensions they engender, are highlighted by efforts to resolve the dilemmas of interpreting sustainability by making resource management an essentially local or regional affair.

Sustainability as 'a local problem'

A wide range of arguments have been put forward for greater local or regional control over environmental resources. Communities living in particular environments are deemed to make better decisions about them, in large part because proximity confers sensitivity to local economic and environmental conditions (Blowers 1993b; Norgaard 1988). Certainly, our analysis of minerals planning highlights the difficulties of mapping universal sustainable development 'blueprints' onto different local contexts: the relationship between economic welfare and the immediate environment differs significantly between Berkshire and Harris (Northcott 2000).[35] Greater local resource self-sufficiency would also reduce the impact of transporting goods over great distances and avoid geopolitical dependence on distant supply locations (Rees 1999). Allowing local and regional communities to escape the strictures of centralised, hierarchical resource development policies has also been deemed to promote more meaningful conceptions of sustainability and justice. There are two aspects to this argument. The first is that achieving greater regional self-sufficiency in aggregates (as advocated by Kellett 1995; Pollock and Henry 1996) would make it more difficult for regions of high aggregates demand, like south-east England, to subsidise growth by importing construction materials, or to export environmental costs to other regions. The second is that giving powers of self-determination to sub-national territorial communities is good in itself, enabling them to make decisions about 'their' environment, quite apart from the outcomes that may arise.

These are important considerations, but there are reasons to doubt the thesis that sustainability at the wider scale could necessarily emerge from the sum of local interpretations. As we argued in Chapter 4, the association between 'planning for sustainability' and 'the local community' is problematic, given that local communities may not necessarily arrive at conceptions of sustainability or justice that would be regarded as defensible in a wider society (Campbell and Marshall 2000; Eckersley 1992; Harvey 1996).[36] While some would favour particularist over universalist approaches to determining what is sustainable, the former are difficult to defend when communities are so obviously not hermetically sealed. Not only do environmental effects transgress political boundaries – the greater visibility of the proposed Jondal superquarry from the municipality on the other side of the fjord is a good illustration – but many interests, with different territorial constituencies, also have a legitimate stake in particular 'local' environments, as is the case with minerals extraction in areas of scenic and ecological value. Nor are communities autonomous in an economic sense: both the circumstances generating the demand for extractive industry and the vulnerabilities of the localities that agree to accept it are driven by economic forces operating at a much broader scale. Freeing communities from hierarchical policy structures does not mean that they will have, or be able to acquire unaided, the agency to make effective use of that freedom (Benton 1993; Low and Gleeson 1999). But in any case, focusing on how much political autonomy or control an individual community might enjoy misses the point that effective organisation and action often depend upon exploiting institutional connections between places and scales. It is significant, for example, that resistance to both the Scottish and Norwegian superquarries proved most successful when local actors were able to establish a broad base of opposition, extending beyond local arenas to embrace wider social coalitions. This is a crucial point. Diminishing the political links between places, even if coupled with the hope that local communities will strive for greater resource self-sufficiency, is unlikely to reduce significantly the way in which modes of production link places together; rather, the effect might be to diminish the space for collective action in the face of the market.

For a variety of reasons, therefore, it would be inappropriate to focus in minerals planning, or in any other sphere, upon local interpretations of sustainability, and it is certainly difficult to see how the political project of strong sustainability – embedding the economy within environmental boundaries – could be promoted through local political and developmental autonomy. These are not entirely new issues:[37] the important point is that it would be misguided to pretend that simply by invoking sustainable development, enduring dilemmas for multi-level government,[38] including the pursuit of justice (Low and Gleeson 1998), could be resolved. While there are practical, institutional and ethical reasons for devolving decision making (King and Stoker 1996), the existence of collective, cross-border concerns makes it necessary to reconcile local priorities with a basis for defensible judgements that can encompass a broader community (D. Smith 1994).

One difficulty in effecting such reconciliation is that translating issues between spatial scales allows different moral claims to slide into and out of focus (Reed and Slaymaker 1993). For example, even if Harris would support a greater number of (wealthier) islanders with the development of a superquarry, some residents felt that this could not eliminate the possibility of environmental, economic and cultural costs being imposed on specific individuals – existing islanders and their children.[39] Yet the importance of bridging shared principles and particular circumstances in promoting sustainability and justice gives some urgency to the task, which can be approached from two directions. First, closer scrutiny of the categories 'universal' and 'local' circumstances dissolves some of the antithesis between them. Our analysis of minerals planning illustrates that it is rarely possible to resolve debates on the basis of principles alone, not least because, in application to particular decisions, different principles may conflict. Recognising this is an advance on believing that consensus can be achieved simply because all communities subscribe to sustainable development. Similarly, responsiveness to context does not necessarily hinge on relativistic principles, and universal principles do not entail uniform outcomes (O. O'Neill 1996). As we have shown, there may be morally relevant reasons for differential treatment. In addition, 'universal' principles in large part derive their veracity, and therefore their ability to travel from one arena to another, from their capacity to inform defensible outcomes in particular contexts. The second direction is to focus on the institutional and political arrangements that might best allow the priorities of different communities, and multiple conceptions of justice, to be debated (see Low 1994). Deliberative and inclusive processes would seem to have much to offer, especially if directed towards cultivating the virtues of tolerance and respect (*ibid.*; Munton 1999). However, their potential remains relatively untapped in the context of cross-border planning and sustainability issues; as we observed in Chapter 4, many experiments have focused on agendas *within* particular localities. Our analysis of minerals planning also injects some notes of caution about the capacity of novel deliberative processes to overcome conflicts where matters of identity and sovereignty are at stake. That said, international trade, environmental concern and political devolution did begin to redefine 'national' resource policies during the 1990s. We turn next to examining how far recent changes to systems of governance in minerals planning are enlarging the political space for deliberating conceptions of sustainability between political communities.

Towards a less hierarchical minerals planning?

Planning systems embody principles of justice, both in the use of planning to pursue particular distributive outcomes and in the rights and values embodied in the institutional arrangements. Planning has traditionally been associated with the pursuit of universal rights to safe, healthy and pleasing local environments (Agyeman 2000; Low 1994; Low and Gleeson 1999). Rights-based frameworks feature (however inadequately) in the scope for public involvement, and the spatial inequalities arising from the siting of minerals development have at least been deliberated within the planning process, as our earlier assessments show. However, action to address these inequalities is significantly influenced by other conceptions of justice, embedded in the framework of property rights (Ellis 1999). As we noted earlier, planning authorities cannot easily change the economic extent of existing mining rights without paying compensation to the holder of planning consent; thus inherited institutions retard efforts to protect and enhance the environment.[40] Compensating communities that are burdened with a disproportionate level of environmentally damaging development 'in the wider public interest' is an idea that has generated much attention in planning debates as a way of attending to distributive injustice.[41] The use of compensation to offset the local social costs of new development is highly problematic, however. Such 'technocratic utilitarianism' (Low and Gleeson 1998: 115) implies that all social and environmental costs can be fully compensated for, that they are essentially 'local' in character, and hence that the provision of compensation can legitimise the decision (Goodin 1989). In the Lingerbay example, however, not all islanders would have taken higher incomes in compensation for threats to the landscape or island culture. Consequently, addressing distributive injustices through compensatory means cannot obviate the need to address matters of procedural justice in the way in which different communities participate in the broader decisions about production and consumption. The question, then, is how far planning allows such participation to take place.

The planning systems of most European countries operate by articulating 'higher-level' principles with flexibility in local application: these higher-level principles increasingly incorporate international and European objectives in national policy imperatives. At the level of practice, too, planning professionals have some scope to bridge individual interests and collective welfare (Forester 1994b) and mediate between different social groups. They may also be required to translate national or regional policies into local decisions. The way in which this translation takes place varies significantly between policy sectors and nation-states. We have contrasted Norwegian minerals planning, in which aggregates extraction is an 'ordinary' category of development, without a national strategic agenda, with minerals planning in England, which channels particular discourses of economic need into local plans. Viewed positively, this strategic framework has given visibility to government judgements about 'need' and 'demand' (Jewell 1995) and has provided some basis for dialogue between planning tiers; however, the channels for debate are restricted in a number of ways. Geological knowledge and economic modelling dominate the construction of minerals policy, with the minerals industry being represented at all levels in the planning hierarchy as a key source of relevant expertise (Beynon *et al.* 2000; Ellis 1999).[42] As we noted earlier, other actors have had difficulty challenging definitions of an 'adequate and steady supply', which constitutes the government's 'dominant strategic line' (Jessop 1997b: 13). Arguably, greater procedural fairness in aggregates planning would have to entail reducing the 'discursive domination' (Healey 1997: 193) of demand projections in the communication between spatial scales. Opening up this 'black box' could expose underlying judgements about producer and consumer sovereignty in determining needs for minerals, as well as the regional and economic interests that benefit from particular conceptions of the 'national good'.[43] One should therefore expect such a step to be contested, and emerging practice also suggests that demoting projections is insufficient in placing cross-border and cross-scale planning issues on a more dialogical plane.

The prospect of urban regions in northern Europe satisfying their aggregates demands from distant superquarries prompted some commentators to call for a European policy on aggregates provision (Wolters and van der Moolen 1999). To date, however, the most significant influence of European policy making on minerals planning has been through environmental regulation. Many of these policies have been implemented through directives, which establish basic, common principles but allow member states flexibility in the means of implementation. The Habitats Directive, for example, has placed significant restrictions on minerals production, and the revised Directive on Environmental Assessment requires member states to introduce mechanisms for considering cross-border impacts and consultation – a good illustration of a 'technical' approach creating political apertures, as we suggested in Chapter 4. Initiatives arising from the European Spatial Development Perspective may also prove beneficial in this regard. Outside the environmental sphere, the European Union has not wished to be seen as usurping control of resource exploitation (Vanderseypen 1998),[44] and the Commission has focused on achieving a 'level playing field' within the single market. In this paradigm, stricter national environmental regulations are allowed provided they do not introduce a 'distortion'; hence the policy sphere is limited, as it is in debates about world trade more generally, to setting orderly rules for competitive exploitation without addressing the scale or flows of economic activity that might arise.

Within the UK, we have seen the further devolution of planning powers: a step that may 'change the ways in which problems are defined and the range of actors whose behaviour influences a solution' (Meadowcroft 1997: 179). At the time of writing, some initial implications for minerals planning can be detected. The Lingerbay superquarry inquiry decision is perhaps the first major illustration of how devolution has reframed the 'national interest' (i.e. of the UK), diminishing the power of such assertions to align actions across national borders. With Scottish Ministers empowered to make the final decision, any benefit that a superquarry in Scotland might create by reducing land-won aggregates extraction in England could be disregarded. One might hope that the greatly expanded space for political deliberation created by the Scottish Parliament will see any scope for cross-border 'spatial fixes' between England and Scotland being opened up to debate, allowing issues of sustainability and justice to be considered in what, hitherto, have been largely closed discussions between state minerals planning officials. In practice, Scottish Ministers chose to amplify the Lingerbay decision by arguing that 'the planning system has no control over the eventual market to be served and that production from Scottish superquarries cannot be directed ... to English markets' (Firth 2000: para 28). This suggests a withdrawal of interest in engaging in policy debates, in distant locations, that influence demands for superquarry material.

The view from England acknowledges the diminished reach of the 'national' planning arrangements and the ethical implications of seeking contributions to domestic aggregates demand from 'imports'. In the draft consultation paper, *Planning for the Supply of Aggregates in England*, the government stated that 'planning policies for aggregates in Wales and Scotland are matters for the devolved administrations in those two countries', hence 'policy for England' cannot 'require, or assume, new or replacement capacity for exports from Scotland or Wales'; or indeed from other countries, such as Ireland and Norway (DETR 2000f: para 3.15). Moreover:

> Active policies of greater reliance on such sources would ... attract the criticism that England is simply 'exporting its environmental damage'. On the other hand, some sources remote from the main English markets may have greater 'environmental capacity' to sustain aggregates production, and be subject to national planning policies tolerant of making provision that includes exports.
>
> *ibid.*

This returns us to one of our central themes, whether planning could prove an effective arena for deliberating and differentiating between the export of unsustainability and the utilisation of distant environmental capacity for development. In practice, the DETR has been keen to follow Scottish Ministers in leaving the cross-border flows of material to 'the market', retreating from any notion of connected public interests.

Within England, changes to the minerals planning regime have emerged alongside new arrangements for regional governance, with central government attributing almost magical properties to an enhanced regional planning framework for overcoming former antagonisms between national and local agendas, as well as between economic development and environmental protection (Murdoch and Tewdwr-Jones 1999). Certainly, central government has conferred greater legitimacy on the substantive policy options available to regional planning bodies, especially to pursue structural fixes that might reduce the need to produce primary aggregates (see, for example, DETR 2000c). A willingness to rethink the role of demand projections can be read as a corollary of concerns to make devolution meaningful, enabled by the blow dealt to the credibility of the 1992 projections by the slump in demand. There have been indications that national-to-local relations may become less closely prescribed by technical planning discourses, with aggregates supply planning moving from a 'predict and provide' mentality to become 'more responsive to local circumstances' (Raynsford 2000). Some form of sustainability evaluation and testing in public are central components of this responsiveness, through which the government hopes that each regional planning body will 'develop and endorse its own objectives for sustainable development in a consensual and inclusive way' (DETR 2000c: para 3.1).

However, it is by no means certain that this flexibility will materialise, or that regional planning will be able to generate consensus, when the underlying issues of minerals planning – need, demand and location – have different conceptions of justice and sustainability at stake. The evidence of the 1980s and 1990s is that consensual, managerial solutions will only take things so far when different coalitions hold distinctive perspectives on whether environmental qualities or certain material demands are ultimately less negotiable. In turn, those interested in divergent substantive conceptions of sustainability may continue to argue instrumentally in favour of particular relations between the scales of decision making. The minerals industry still favours a broadly top-down planning framework in which projections – albeit assuming static rather than growing demand – guide local planning authorities to allocate sufficient land for aggregates provision (Pollock 1999). The CPRE (2000a: 2) and other environmental groups see the street running the other way, seeking 'greater local discretion' to set 'quarrying levels consistent with environmental capacity'. Over the top of this jostling for position, the government has already fired a number of warning shots. It 'will not fail to intervene' (DETR 2000f: para 4.23) where regional or local approaches seek 'consciously [to restrict] the total supply from all types of aggregates below a reasonable and justified expectation of demand' or, *inter alia*, '[seek] to secure amounts of permitted reserves of primary aggregates in excess of those which are needed to meet justifiable and reasonable requirements' (*ibid.*: para 6.2). This concern with justification might herald a more deliberative approach to minerals planning in which even policy cores are open to debate. A political willingness to grapple with such fundamental issues would make it even less likely that regional planning will provide a space for conflict resolution, rather than 'another site where ... conflict can take place' (Jessop 1997a: 122).

Conclusions

Significant changes have undoubtedly taken place in British minerals planning since the 1970s. Steady improvement in the quality of environmental management and restoration at many sites has been joined by a marked change in the language of government policy, now supportive of structural fixes, with a key policy goal becoming 'to reduce the demand for aggregates by minimising the waste of construction materials and maximising the use that is made of alternatives to primary aggregates so that less is dug from the ground' (DETR 2000f: para 7).[45] While this policy approach was being put forward by environmental groups as long ago as the 1970s (Searle 1975), conceptions of sustainable development seem to have provided a legitimising discourse for the merging of problem and policy streams (Kingdon 1995) in minerals planning. In terms of the questions with which we started this chapter, this shift could help to smooth relations between the priorities of different scales of government by reducing pressures to identify new extraction sites.

Such a process is unlikely to appear smooth on the ground, however, because national shifts in policy agendas arise from minerals planning conflicts in a large number of different localities. Although eco-modernist conceptions of sustainability often suggest that dematerialising production has an innate 'win–win' economic logic, it has taken the pressure of the decreasing availability of 'suitable locations' for minerals production to encourage the government and industry to contemplate alternatives to the traditional supply-led approach. Powerful opposition to a Lingerbay superquarry challenged the legitimacy of seeking aggregates supplies from Scottish superquarries and, *inter alia*, the idea of an ever-extending resource horizon. Political changes to the spatial structures of governance have also had an impact. Devolution in the UK reduced the political scope to reconcile the economic demands of south-east England with a UK-wide 'environmental capacity'. Economic conditions are also important, with the aggregates industry itself being generally cautious about the commercial opportunities for coastal superquarries. High transport costs continue to ensure that most aggregates demand is met from relatively local sources. Altogether, these processes have 'narrowed the options' (Blowers and Leroy 1994: 198) for government, hastening the need to confront the issue of which minerals demands are 'essential' and which environments are 'critical'. Whatever its deficiencies as an arena for public deliberation, it is clear that the arrangements for public and interest group engagement provided by the planning system have created apertures for searching critiques of broader economic processes.

One should not rush to proclaim these incipient changes as evidence of policy learning, however, or of a durable realignment between government, industry and environmental groups. Certainly, most major minerals companies have rethought their economic interests and are now actively engaged in the production of recycled and secondary aggregates (ENDS 1998c); this demonstrable market feasibility has doubtless emboldened the government, but medium-term economic priorities also underpin this nascent discourse coalition (Hajer 1995). Falling UK demand for primary aggregates in the 1990s caused prices to collapse. For the major aggregates companies at least, planning policies that constrain the development of new quarrying capacity and raise prices fit with their own economic strategies in such circumstances.[46] If there seemed to be some convergence of interests around policy outcomes by the start of the twenty-first century, it might have obscured potential conflicts over policy arrangements. Although the QPA has joined environmental groups in criticising the traditional system of demand projections, regarding the present models as 'discredited and pointless' (Pollock 1999: 14), we have noted that the two sides do not share views on what should replace it. A greater degree of sub-national flexibility in the regulation of aggregates production in the UK seemed inevitable, given political devolution. At the same time, the fact that the 'objects of governance' have become less problematic (because likely aggregates demand is recognised to be lower than ten years ago) may *allow* the 'modes of governance' (Jessop 1997a: 105) to become more flexible. The precise scope for flexibility at local, regional and national planning levels is being keenly debated at the time of writing, but, given the diverse and conflicting political principles being articulated, it might be naive to anticipate a stable consensus on the role of regional planning.

As we have shown, theoretical reflection on sustainable development has not generated consistent guidance on governance arrangements, beyond stressing the ecological connections between actions at all scales, since different conceptions of scale are bound up with different conceptions of sustainability. However, if we accept the thesis that the politics of sustainability entails multiple, overlapping fields of political authority, then one can see why seductively simple prescriptions, which concentrate authority on a single political tier, are unlikely ever to be satisfactory. In many respects, British minerals planning reflects this situation, with much attention being devoted to the evolving vertical relations between local, regional and national government. That said, relatively little consideration is being given to the horizontal connections *between* regions or countries, based around actual material flows. Elsewhere in Europe, the foundations of cross-border minerals planning are being laid, such as that between North Rhine–Westphalia (Germany) and the Netherlands (see Ellerbrock 1999). In the UK, however, the emphasis on a streamlined approach to regional planning, the regional 'ownership' of planning strategies and regional competitiveness risks driving out concerns about sustainability and justice that cross national and regional borders. There are few signs yet in minerals planning of a concern to place material flows in a wider ethical framework, 'formalising portions of … distant ecological footprints' through 'ecologically balanced and socially fair exchanges' (W. Rees 1999: 121).

Finally, it seems doubtful that the hesitant shift towards a more ecologically modern approach to minerals planning will eliminate locational conflicts. Large volumes of aggregates will still be required to meet needs that cannot (yet) be met by alternative materials, and even greatly improved site management would not eliminate concerns that particular siting decisions are unjust (Blowers 1999). In our assessment of the interactions between conceptions of sustainability and minerals planning, we have identified powerful trends towards the concentration of remaining production on existing quarrying areas, reproducing the geography of minerals production over time and shaping the present and future fate of quarrying communities. Processes of peripheralisation evidently connect intra- and inter-generational equity in quite a forceful way. If we are dissatisfied with the idea that scattering quarrying across all landscapes (albeit perhaps in smaller, better-managed sites) represents a desirable or feasible solution, then the question becomes how far concentrating extraction on existing sites can be made environmentally sustainable and socially just. The acquiescence of current or potential quarrying locations is insufficient, morally, to absolve the responsibilities of those that may benefit from the resources, and in any case it appears to be diminishing. Beyond achieving incremental improvements to quarrying management and restoration, however, the role of planning appears limited. As a regulatory system, it makes little impact upon the underlying structural inequalities in economic power that are implicated in uneven development, or on the system of property rights that distributes the benefits from resources. The most fruitful agenda lies in achieving greater procedural fairness in planning, allowing quarrying communities to voice their concerns in local, national and international planning arenas and contribute to deliberations over the production and not just the distribution of environmental burdens (Faber 1998). The relative openness of planning, and the unavoidable conjunction of substantive and procedural notions of justice in planning for sustainable development, will ensure that it remains a vital institutional setting for working through the conceptual tensions in practical settings. One might at least hope, if devolution slows down the process of policy formulation (Gleeson 2000), that this will allow these ethical questions to be debated. We consider the scope for this, and for other developments, in our concluding chapter.

8

CONCLUSIONS
AND REFLECTIONS

Unfortunately for the compromisers, many land use controversies cannot be reconciled. The politician seeking 'balance' is likely to satisfy nobody.

Caldwell and Schrader-Frechette 1993: 14

How seldom is it that theories stand the wear and tear of practice!

Trollope, in *Thackeray* (1879)

Introduction

We have been concerned in this book with the interface between an idea – sustainable development – and a set of institutions regulating the use and development of land. This interface has been a complex and dynamic one, not least because of the intricacies of interpreting sustainable development, particularly in the context of land-use change. In Chapter 3, we argued that moving from the broad consensual concept towards operational principles must entail important judgements of value: the result is that different groups espouse divergent conceptions of what it means for development to be sustainable. Particular interpretations may be mobilised in the policy process by (ostensibly neutral) techniques, different forms of participation and appeals to 'integration' of environmental, economic and social goals. Indeed, our discussions of policy and practice, conceptual issues and 'instruments' for sustainable development in the first half of the book all suggested that the idea of implementing, or operationalising, some predefined and widely agreed concept is profoundly misleading. Later chapters on transport, nature conservation and minerals planning showed, in specific empirical contexts, how divergent interpretations of sustainability have vied for supremacy in different planning forums, and explored the extent to which new thinking has impacted upon wider political processes. Our objective in this final chapter is to draw together and reflect upon our main arguments and findings. We summarise key developments and revisit our original questions about the fate of a challenging new idea in an active policy domain. In

158

the light of this experience, we offer some thoughts on what a land-use planning system *ought* to be if it is to engage in any meaningful way with conceptions of sustainable development. However, we acknowledge that substantive issues remain unresolved.

An active interface

One firm conclusion is that the concept of sustainable development has generated much activity in land-use planning and related spheres: we would concur with Wilson (1998: 49) that 'the extent of experimentation with radical new principles, policy instruments and techniques has been remarkable'. The emergent concept resonated widely with the structure and concerns of land-use regulation, while a belief that planning would be vital in delivering sustainability came as a welcome counterweight to the neo-liberal agendas of the 1980s.

One notable result of this engagement has been a marked growth in the collection of data, demanded at all levels to inform appraisals, audits and indicators of various kinds. Not only the scale of this activity but also the information deemed to be relevant (on habitats, air quality and recycled resources, for example) might be seen as indicative of broader shifts in policy (Hall 1993). In parallel, as we indicated in Chapter 4, many planning bodies have experimented with new ways of engaging the public, reflecting the participatory ethos of sustainable development and disillusion with traditional approaches. Policy 'software' (Weale 1992: 100) has also developed, in the form of processes and networks attempting to link and co-ordinate policy sectors, social goals and spatial scales. Some, such as the ROM projects in the Netherlands and local air-quality management groups in the UK, are more or less formally required; others, including the networks involved in local biodiversity action plans and Local Agenda 21, are informal. 'Invisible colleges' can also be identified, pursuing shared values and policy interests across national boundaries (Crane 1972; Gilbert *et al.* 1996; White 1993): the 'new realism' in transport was partly advanced in this way, as we saw in Chapter 5. Both the gathering of information and attempts to integrate and co-operate can stimulate learning and help to reframe important issues, although neither leads necessarily or quickly to policy change. However, perhaps the most visible effect of engagement with sustainable development has been the impact on plan making itself. At least at a rhetorical level, there has been widespread, if geographically uneven, take-up of the concept in land-use and spatial planning documents of all kinds.

Significantly, these important developments have taken place, in the UK at least, with little change to the legal and institutional basis of land-use planning. The introduction of a plan-led system in 1990 had implications for, but was not itself a product of, new thinking about sustainable development, and the regions underpinning the spatial planning initiatives of the late 1990s reflected traditional administrative boundaries. Devolution redistributed planning powers but scarcely redefined the traditional remit of land-use regulation. Refitting essentially the same system with modified objectives has produced difficulties and inconsistencies, to which we return below. Nevertheless, our analysis has confirmed the significance of land-use planning as an important arena in which the abstract concept of sustainable development is rendered material. Whether concern for sustainability has delivered anything genuinely novel, however – one of the key questions in our introductory chapter – is another matter altogether.

Making a difference?

If we were to look for clear gains – consistent delivery of economic and social objectives alongside effective environmental protection, or rapid and lasting changes in trends and policies that have been widely regarded as unsustainable – then we would have to conclude that, in most areas, the impact of a great deal of activity and fine rhetoric has been limited. In spite of an outward concern for sustainability in all of the sectors that we examined, traffic-generating development persists, provisions for nature conservation struggle to keep pace with pressures for growth, and technical and spatial fixes still have a major role in defining 'sustainable' minerals production. Planning policies seeking restraint in economic hotspots are frowned upon as a threat to prosperity, while in less prosperous areas regional growth coalitions persist in seeing new and improved infrastructure as essential to economic regeneration and development. Real change remains quite difficult to discern.

In Chapter 2, we postulated two kinds of explanation for the gap between aspirations and outcomes. One was essentially to do with timing. In this view, the inclusion of sustainability principles in national and local planning policies – even in vague terms – is only the first stage in 'discourse institutionalization' (Hajer 1995: 61), preparing the ground for more substantive policy change in due course. To use an earlier analogy (Owens 1994: 451), the 'Trojan horse' of sustainable development, concealing radical potential changes, has rolled through the policy gate. The alternative explanation had more to do with underlying structures. As the more far-reaching implications of sustainability have become apparent, responses have been put in place to neutralise any serious challenge to the dominant paradigms of growth and development. Public and professional enthusiasm might still be encouraged (even seen as a helpful distraction) provided that it is channelled in such a way that nothing fundamental is likely to change. In this view, the Trojan horse is immobilised and emptied of its radical contents. In reality, what we have found is something lying between these extremes, requiring a more nuanced explanation: radical conceptions of sustainable development have not triumphed in land-use planning, but neither has the discourse been entirely without power.

The 'greening' of development

The impact of the broad concept has been in evidence in a number of interesting ways. Our analysis suggests that weaker conceptions of sustainability, which emphasise balance or trade-off, and 'Panglossian' ones, which underplay conflict, may still end up promoting greener growth rather than simply defending the *status quo*. Such incremental 'greening' is in evidence in many forms, some more convincing than others. When development proceeds it may be subject to additional constraints, as in two of the cases described at the opening of this book: Cairngorm will have its funicular railway, but visitor movements on the summit will be restrained to minimise ecological damage; and a 'green travel plan' helped to win planning consent for Vodafone's headquarters on a green field site at Newbury. Technical and spatial fixes can improve efficiencies and reduce impacts, as we have shown for the transport and minerals sectors, and interests may be redefined as this potential comes to be recognised. So, for example, congestion became part of the business case for an integrated transport policy, and planning constraints have helped to stimulate the market for secondary and recycled aggregates, offering economic opportunities as well as environmental gains. There is also growing acceptance that environmental losses should be 'compensated' for, for example by creating new habitats when existing ones are damaged or destroyed. And in a few cases, there have been at least temporary 'wins' for the environment, outcomes in which (it might be said) critical environmental capital has been protected: cancellation of the Salisbury by-pass – another of our opening examples – and refusal of consent for the Lingerbay superquarry (see Chapter 7) are prominent illustrations.[1] Through procedures such as environmental assessment, developers have been expected to address the immediate and distant effects of their projects, and even technocratic approaches have helped to stimulate change where they provide a space for opposition and dialogue. In numerous ways, therefore, attention to sustainable development within the planning process has helped to identify new routes out of environment and development conflicts, or to rediscover and legitimise older ones; the overall result, while not radical in any normally understood sense, is that outcomes are less environmentally damaging than they might otherwise have been.

Reframing problems and policies

In itself, however, such an evolution would not be remarkable. Incremental change is the least we could expect thirty years on from the environmental revolution of the 1960s, and it might even be argued that sustainable development should be read as a product rather than a driver of social and political change over this period. But our analysis does not stop there: it suggests that promotion and adoption of the concept of sustainability has had more subtle and potentially more important effects. Appeals to sustainable development have contributed to a reframing of issues in all of the sectors that we examined, sometimes justifying the kinds of incremental improvement outlined above but in other cases pointing towards more radical shifts of policy. Significantly, land-use planning has almost always provided a vital institutional space for the ideas, argument and persuasion that are conducive to such reframing. When change has occurred, however, it has not normally been the outcome of a deliberative and inclusive process arriving at some consensual interpretation of sustainable development. Rather, the apertures for public involvement have been skilfully exploited by coalitions of local and non-local actors to challenge prevailing ideologies and take new storylines closer to the heart of decision making. In this process, intense public concern about individual projects and plans has often been linked to generic critiques of policy by a broadly environmental coalition.

Such linkages have exposed a dialectic between spatial fixes on the one hand – the traditional remit of planning – and structural change on the other, requiring modification to prevailing patterns of production and consumption. The dilemma is amplified because tensions between growth and 'amenity' have often proved resistant to ecologically modern, 'win–win' solutions: instead, they have been recast as fundamental disagreements about the nature of sustainable development. In some cases, the outcome has been a perceptible shift towards – although certainly not a triumph for – environment-led conceptions of sustainability. This trend is exemplified by the 'new realism' in transport policy, adoption of the strategy of 'plan, monitor and manage' in land provision for housing, and a denting of the demand-led minerals planning system. In all of these examples 'policy learning', if it can be said to have occurred, has been a process stimulated by conflict played out within the planning system. We can conclude, then, that even if it has delivered some tangible benefits, the concept of sustainable development has scarcely fulfilled its much vaunted reconciliatory potential. Rather, the struggle for different conceptions of sustainability has forced more detailed scrutiny of dominant modes of social and economic progress, and in some circumstances there have been modest, but not insignificant, changes as a result.

Resisting change

It is nothing new, as we argued in the introduction to this book, for land-use conflicts to raise questions about growth and social purpose. What is significant is that attempts to put sustainable development into practice have reminded us how profound such questions are, and have re-established fruitful terrain for debate. In conflicts over roads and minerals extraction, for example, we have seen how conceptions of sustainability in which it might be regarded as obligatory to protect environments confront powerful assumptions about property rights, growth, the nature of well-being and the status of consumer preferences. Scrutiny of such assumptions, and of the policies and development trends that flow from them, might be regarded as healthy and welcome: after all, dialogue is precisely what many commentators have called for in the context of sustainable development. But while some groups actively seek to prise open 'black boxes', it is only to be expected that those who see benefit in maintaining the *status quo* will make strenuous efforts to deflect such examination.

One way of resisting challenge is by emphasising and reinforcing the legal and institutional boundaries of the planning system, ensuring, for example, that policies are not permitted to stray beyond their 'land-use' remit, or arguing for a narrow interpretation of what is 'material' to a planning decision. Perhaps the most powerful institutional constraint (in the UK at least) is that a presumption in favour of development lingers on in spirit in the planning system, if no longer in the letter of planning policy guidance. It is there in a deference to (and particular interpretation of) the rights of property, and in a clear conviction, evident in the language and tone of policy guidance, that growth in general must be a 'good thing'.[2] Such a presumption is clearly incompatible with environment-led conceptions of sustainability and sits uncomfortably even with conceptions in which the different components are to be given equal weight.

Of course, institutional boundaries need not be wholly impervious or inflexible. After all, they are consciously constructed: indeed, the expectations of planning as an instrument of sustainable development, even on the part of governments, strongly suggest that changes will be required. Still, it may be convenient to maintain boundaries if it means that difficult issues, such as questions of 'need' or environmental capacity, disappear over the 'edge' of the planning system without reverberating elsewhere. The effect can be to shelter the merits of development proposals from scrutiny and restrict the range of potential solutions to land-use conflicts to those with a spatial dimension. With its boundaries tightly defined, planning is left to find 'suitable locations' (Cowell and Owens 1998: 797) and to pursue mitigation, sometimes taking the least politically sensitive escape route at the expense of environmental or social justice. Nor does this experience seem to be a product of the particular institutional framework in the UK: elsewhere – in environmental planning in the Netherlands, for example – it seems that the effect of applying concepts of sustainability has been a series of incremental improvements rather than radical change.

However, the boundaries of the planning system are not the only – or even the most important – defence against potential challenge: they are powerfully reinforced by ensuring that the wider political context is inhospitable to radical conceptions of sustainability. A review of post-Brundtland developments suggests that environment-led interpretations, initially acceptable if only on 'past deficit' grounds, came to be seen as a threat to core aspects of political economy. The response from corporate interests and governments was not to reject sustainable development – the Trojan horse was already through the gate – but to capture and mould the idea, exercising power in a variety of ways to produce their own conceptions in which growth and competitiveness remained at the core. Thus the shift towards 'Panglossian' interpretations in official policy discourse in the UK can be seen as one in which dominant political ideologies reasserted themselves: powerful actors learned not so much how to become sustainable (in some rationalistic sense of seeing the error of their unsustainable ways) but how to adapt conceptions of sustainable development to their favoured goals and priorities. This narrows what might be achieved by land-use planning, which, at least in any given time frame and location, can deliver only those conceptions of sustainability that are admissible in the wider system. Even a radical new remit, such as a key role in reducing the need to travel, can be rendered ineffective by a countervailing ideology, in this case one that associates growth and competitiveness with increasing (and relatively cheap) mobility.[3] Radical social formulations are likely to fare little better, despite an increasing rhetorical emphasis on the social dimension of sustainability. We noted in Chapter 2, for example, a tendency (at least initially) for references to social justice to be deleted from development plans, and that even when a social dimension is explicit, planners are expected to alleviate the *symptoms* of injustice through land-use measures. Typically, these can do little to tackle the root causes of the problem, although they sometimes lead to poorer environmental outcomes. So, for example, local authorities may be forced by central government to allocate sufficient land to meet projected demand for housing on the grounds that a shortfall will drive up prices and increase homelessness. However, they are powerless to deal with low pay, the loss of public sector housing or the wider economic forces that leave houses empty and unsaleable in some regions while placing others under extreme development pressure.

Opposing forces

Overall, then, it would seem premature to claim that the concept of sustainable development, as interpreted in the planning process, has laid firm foundations for substantive longer-term change. To borrow Heclo's (1974) terms, while the idea has forced a great deal of *puzzling*, it has scarcely diminished the role of *powering* in the policy process. We *do* see evidence of the power of argument and persuasion, and the use of every available opportunity (including formal procedures of various kinds) by groups seeking to promote transformative conceptions of sustainable development, such as those based on respect for environmental limits. When growth is manifest in 'locally unwanted land uses', these interests may be backed by the (often considerable) political power of concentrated opposition in particular localities. We also see the exercise of corporate (economic and political) power in these contexts, shaping the way in which deliberation takes place, promoting particular forms of rationality and mobilising less threatening conceptions of sustainability. Engagement between these coalitions may lead (even through conflict) to a process of learning, of reframing problems and identifying outcomes that satisfy multiple objectives – outcomes that are 'more sustainable' in an ecologically modern sense. But many land-use conflicts continue to defy such resolution, confirming our suggestion (in Chapter 1) that it is harder to be 'modern' when non-material environmental values are at stake. We might see the planning system, in such circumstances, as a front along which divergent world views have come into conflict, sometimes advancing a little in a more radical direction, more often retreating behind the certainties of the *status quo*. In the cases we examined, the economic priorities of growth and competitiveness could still be said broadly to structure the 'reality' of land and development issues. In none of them, however, have prevailing structures of power entirely defused the challenge of sustainability or rendered it wholly impotent. The very fact that even traditionally dominant interests feel the need to justify their position as 'sustainable' – and sometimes to shift it in modest but not insignificant ways – is an indication that the boundaries of what is thinkable have been extended. Even when subject to capture, ideas may continue to have an impact on policies, and – perhaps especially in situations of confrontation – reasoned argument may be a stronger form of power than Flyvbjerg (1998) suggests.

After sustainability

Our argument so far suggests that a radical, transformative role for the concept of sustainable development has been constrained both by the institutional remit of the planning system and by the wider social and economic context within which it is embedded. But we have also suggested that the structures are not rigid, and that apertures for 'public involvement' have provided an important space within which different conceptions of sustainable development have been contested and refined. Although this process has not generally (perhaps not ever) been a consensual one, it has been productive in generating ideas, argument and debate, and in some cases in reframing important issues and leading to noticeable changes in policy.

These arguments point to one way in which systems of land-use planning might develop. That they can be guided by no single consensual conception of sustainable development is inevitable and is not what matters: what does matter is that a collapse into relativism is resisted and a social capacity to contend interpretations maintained. The dialogical aspects of the planning process might be nurtured, recognising that deliberation takes place not only in designated 'deliberative forums' (indeed, that many such forums may never be truly deliberative) but is ongoing in the essentially conflictual procedures of policy formulation. It would be an easy step from here to conclude on a positive note: that we should actively encourage planning to become a forum for debate about 'how we wish to live', as well as being a system with more specific instrumental roles like promoting nature conservation, reducing the need to travel and mitigating the impacts of developments. Such a conclusion would follow the broad logic of our own argument and would be in line with much current thinking about sustainable development and planning. But it would be too simple, for we have also identified substantial obstacles, some presented by the traditional remit of planning systems like that of the UK, and some that pose genuine intellectual problems. Certainly, before planning could hope to fulfil the deliberative role that many envisage, these difficult issues will need to be confronted and some fundamental changes made.

One important dilemma – it emerges clearly from many of the issues discussed in this book – is that while there are good reasons to promote a deliberative and argumentative role for planning, rather than the functional one of *implementing* some singular notion of sustainable development, it is important to ensure that fundamental principles of environmental and social justice are upheld – principles that must themselves, over time and in a wider forum, be the subject of argument and debate. This would present particular difficulties if we were to insist that planning must be a neutral arena, in which different conceptions of sustainable development can be contested but which would give no lead or set no bounds to the debate. Such a model need not detain us, however, since planning never has been and never could be neutral: even in its minimalist form (of ensuring that markets can function) it is implicitly grounded in certain conceptions of the good. And the argument throughout this book suggests that if planning is to concern itself with the normative claims of sustainable development, it cannot avoid adopting positions in relation to different moral claims. This does not eliminate space for dialogue: as we have shown, even when clear principles are adopted, they tend to under-determine action and must of necessity be interpreted in particular contexts.

Environmental justice, for example, might require that planning defend conceptions of sustainability in which certain obligations to protect the environment are acknowledged. This implies, on any normal planning timescale, that there must be limits to the capacity of a given area to accommodate further development (on this we concur with the House of Lords Select Committee on Sustainable Development 1995). Following our argument in Chapter 3, the obligations to be recognised may demand a level of protection beyond that which could be justified in terms of maximising

166

preference satisfaction. However, accepting such a principle would certainly under-determine how capacities or limits should actually be defined, or what actions or policies would be appropriate when there are conflicts of obligation. So here, perhaps, we have a space for dialogue, for arriving at informed judgements about different claims in such a way that 'the intersubjective virtues and standards of argument and criticism apply' (Sagoff 1988: 43). The process of 'informing' will almost certainly include data from a range of sources, and appropriate audit and appraisal techniques, as long as a crude technical rationality is avoided. In such a scheme, professional planners would not aspire or pretend to be neutral facilitators of dialogue: their role would be more akin to that of expert participant, working with informed citizens and policy makers to determine in a given context what might be deemed 'sustainable'.

However, dialogue can hardly be meaningful if it is consistently curtailed along one axis. It would still be difficult to reach informed judgements about what is sustainable if the merits of particular forms of development remained closed to scrutiny. Yet to permit such scrutiny within the planning system – in effect to interrogate the social purpose behind particular development trajectories – would entail changes to the legal and institutional framework that would almost certainly meet well-organised resistance. There is, then, a structural difficulty to the scheme outlined above, certainly for planning systems like that of the UK. But there is also an intellectual one: if institutional barriers need not be insurmountable, where *should* the bounds of the planning system lie? Dialogue may be a good way of dealing with purely local issues, but in the context of sustainability these are probably rare, because many processes shaping economic and environmental change operate at a much broader spatial scale. Our own argument (in Chapters 4 and 7) would certainly challenge what Marvin and Guy (1997: 311) have called the 'new localism' – the belief that sustainable development is best interpreted and pursued at the level of local communities – for even if they are constrained by certain universal principles, it is far from clear how the sum of 'local sustainabilities' could add up to a sustainable whole. At the other extreme, while it seems vital to be able to question the 'monopoly of social purpose' behind many development pressures, it would hardly be practicable to subject every market trend or government policy to profound scrutiny in plan making or development control. One way forward would be to build upon (and resist pressures to curtail or 'streamline') the demonstrable strength of the planning system in opening up terrain for debate, particularly when there are real incompatibities between different social claims. This would recognise that while not all decisions can be local, political issues of wider significance are often crystallised and made material in local contexts. One thing that is clear is that we need to move beyond the broad consensus that there should be 'deliberation' in planning to work out what, and at what level, might properly and productively be deliberated.

This leaves us with the vexed question of who, exactly, should participate in this process. We have shown, in various contexts, that making the planning system more pluralistic has undoubtedly been important in the wider deliberation of policy objectives and has generally been a force for good. But the problems of more inclusive engagement are substantial, and it would certainly be premature to conclude that sustainable development, as interpreted in the planning system, has fulfilled its more ambitious, participatory promise. Even if we can provide answers to some of the bigger questions posed above, planning may not provide a ready-made context in which inclusive deliberation about sustainable development will simply happen. If meaningful engagement is to be extended, the system may first need to become a *beneficiary* of a new political culture in which people see themselves as citizens, 'willing, able and equipped to have an influence in public life' (Advisory Group on Citizenship 1998: 7). Quite how this is to be brought about, and how even then we might bridge the particular and the universal in conceptions of sustainability, must remain the subjects of more rigorous thought and analysis.

There is, then, much work to be done. But to show that fundamental questions are inescapable is not to suggest that the concept of sustainable development is vacuous, or that the argument and debate that it has generated have been futile. On the contrary, we have shown throughout this book how the general concept has provided a starting point for a series of important debates. As we argued at the outset, decisions about the use of land demand the interpretation of sustainability in ways that have visible and material implications for social and economic activity. This unavoidable yet contentious task has made planning an important arena for conflict, and the debate generated contributes in a modest way to the wider deliberation of social purpose. Although some might see the original concept of sustainable development, subject as it has been to capture and manipulation, as something of an empty husk, we would conclude on a more constructive note. Like other broad but important concepts, that of sustainable development has performed (and could continue to perform) a vital task. It has done so not by resolving all planning conflicts but by focusing attention on the claims of environmental integrity, social justice and a dignified quality of life, and on the substantial moral and political task of adjudicating between claims that cannot always be happily reconciled. That the process is taxing is to be expected, and welcomed. As others have recognised, 'the path of great principles is marked through history by trouble, anxiety and conflict' (Bagehot 1856: 287).

NOTES

1 Old conflicts and new ideas

1 Vodafone, the UK's largest mobile telephone group, proposed new headquarters for 3,400 staff on a site outside the town of Newbury. West Berkshire Council granted planning permission in April 1999.

2 Project promoters included the developers (the Cairngorm Chairlift Company) and Highlands and Islands Enterprise. The application, submitted in August 1994, covered a funicular railway to replace an existing chairlift, and a visitors' centre close to the summit of Cairngorm. Although planning permission was granted in 1997, protection of the fragile arctic flora and other wildlife requires that visitors be confined to the centre.

3 The original inquiry was in 1993. The road would have been elevated 5 to 7 metres above the valley floor, threatening views of Salisbury and its cathedral spire in a pastoral setting. These were, according to the Countryside Commission (1997: 1), 'uniquely valuable and of importance to our national heritage'. The scheme was cancelled in 1997 in the 'accelerated review' of certain road projects (see Chapter 5), and a study to consider alternative solutions to the area's traffic problems was promised.

4 The guidance, intended to provide a framework for the preparation of local authority development plans across the region over a fifteen- to twenty-year period, identified the scale and distribution of provision for new housing and set out priorities for environment and economic development as well as for more specific issues such as transport, minerals and waste. The public examination was held in 1999.

5 The Brundtland Report provides the most widely cited, and broadly accepted, definition of sustainable development: 'development that meets the needs of the present generation without compromising the ability of future generations to meet their needs' (WCED 1987: 40).

6 Ecological modernisation has increasingly been used as a framework for describing and analysing changes in approaches to environmental issues and policies (Christoff 1996; Drysek 1997; Hajer 1995; Mol 1996; Weale 1992). It also refers to the normative beliefs of those who seek to promote such developments, particularly those embracing proactive and integrative approaches.

7 We have found it unhelpful to adopt the distinction sometimes made between 'sustainability', with an environmental focus, and 'sustainable development' as being more overtly concerned with the meaning of progress and its distributive effects. We argue that that definition is itself part of the political process, woven into land-use conflicts, and we use the terms interchangeably in this book.

8 Rawls (1972: 5) writes:

it seems natural to think of the concept of justice as distinct from the various conceptions of justice and as being specified by the role which these different sets of principles, these different conceptions, have in common.

9 'Learning' is central to cognitive perspectives on the policy process and is the subject of a growing multi-disciplinary literature (see, for example, Hall 1993; Jachtenfuchs 1996; Jachtenfuchs and Huber 1993; May 1992; Parson and Clark 1997). This often draws inspiration from Heclo (1974: 306), for whom learning was 'a relatively enduring alteration in behaviour that results from experience'. Sabatier (1998) builds on this definition to include information as a stimulus, and thought, as well as behaviour, as something that might be altered. Some authors prefer a less rationalist, more discursive, conception of learning: Hajer (1995) suggests that policy change occurs when issues are 're-framed', or selected, organised and interpreted in new ways (Rein and Schön 1991). Conditions previously assumed to be inevitable (such as the traffic growth and dispersal of land uses that we discuss in Chapter 5) can be seen afresh as 'problems', meriting political attention.

10 Change is not restricted to such revolutions; as Wright (1985) has argued, new accumulative cycles continually demand transformations of both urban and rural environments.

11 This is partly due to the nature of primary activities (such as quarrying) but also because land is not so much consumed in the process of development as transformed; what matters is what is gained or lost in the transformation.

12 'Spatial planning' has acquired a more specific meaning in some instances, referring to the strategic co-ordination of sectoral activities with territorial implications – economic development, transport, health, education, and so on. We discuss some applications of this concept in Chapter 2.

13 It should be noted that the physical development of land is only one component (perhaps not even a necessary one) of 'development' in a more general sense, which must include economic and social dimensions.

14 Either (in neo-classical economic theory) because it achieves an efficient allocation of resources, in the equilibrium condition of Pareto optimality, or (in the Austrian tradition) because it provides the best conditions for co-ordinating the diverse and dynamic plans of many different actors (for further discussion, see J. O'Neill 1998). Interestingly, Hayek, the arch-critic of centralised economic and social planning within the Austrian School, did accept a case for town planning 'to supplement and assist the market' but not to 'put central direction in its place' (Hayek 1960: 350, quoted in Lai 1999: 1571). It was acceptable to make the unit of competition larger than that of the individually owned property, but Hayek was opposed to town planning at any level above that of the district. (For an interesting discussion, see Lai 1999).

15 McAuslan's classic critique might be seen in this light. Planning law, represented as a neutral and objective set of rules, 'a golden metawand', is in practice no such thing: 'The words themselves, the way they are put together and have been interpreted, embody ideologies and beliefs about power and society' (McAuslan 1979: 2). Similarly, Grove-White (1991) argues that planning law not only provides regulatory constraint to possible courses of action but, through its discourses and idioms, and the particular opportunity structures it provides, it has also exerted a wider cultural framing on the forms in which environmental tensions have been conceptualised and have found public expression in the UK. Particularly important in this context has been determination in the courts of what constitutes a 'material consideration' in decisions about land use (Grant 1998).

2 Rhetoric, policy and practice: sustainable development as a planning issue

1 Many other European policies, including those on pollution control, environmental assessment and aspects of urban and regional development, were also clearly relevant in this context.

2 Again, reflecting a more general tendency for new concepts to provide containers for older plans and projects. As van der Straaten and Ugelow (1994: 126) observe of the plan: 'Although ideas of sustainable development led the discussion, ... the material hailed as innovative had actually appeared years before in several other reports'.

3 Championed by Vice President Al Gore. Addressing the National League of Cities Annual Conference in Kansas City in December 1998, Gore criticised 'neon nightmares' of urban sprawl and called for carefully focused federal government incentives for 'smarter growth', such as aid for transport innovations in Portland, Oregon (which we discuss in Chapter 5) (*San Jose Mercury News*, 5 December 1998).

4 In Section 31 of the Act. The Development Plan Regulations 1991 require local authorities to have regard to environmental, social and economic considerations when preparing structure plans and unitary development plans.

5 For example, in an address by Planning Minister Sir George Young to the Town and Country Planning Summer School on 11 September 1991, and in speeches by Michael Heseltine, then Secretary of State for the Environment, at the *Surveyor* conference on 23 January 1992 and at the Government Conference on Regional Planning Guidance for the South East later in the same year.

6 While emphasising the role of the planning system in the supply of land for development, the CBI and RICS recognised that 'the system must meet broader public concerns about environmental protection and social well-being' (1992: 6). Sustainable development was endorsed, if only obliquely, through references to planning policy guidance.

7 This is not just a matter of land-use changes in any one location having multiple environmental impacts; it is also a relationship in which the activities of a given population – that of a city, for example – have implications for resource consumption and land use on a much wider scale. Land use is central to the concept of the 'ecological footprint' – the total area of land and water (anywhere on Earth) required to provide resources for, and assimilate the wastes from, such a population (Borgström Hansson and Wackernagel 1999; Wackernagel and Rees 1996; Wackernagel *et al.* 1999). The idea that each human being occupies an 'environmental space' is broadly similar, with global environmental capacities to yield certain services or assimilate wastes being divided by global population to calculate notional individual entitlements (Carley and Spapens 1998; Friends of the Earth Europe 1995).

8 The remit of British planning authorities in relation to emissions (mainly from stationary industrial sources) statutorily regulated by other bodies has been seen by successive governments as locational (restricting development of polluting activities in the vicinity of 'sensitive' land uses, or *vice versa*). It does not extend to the imposition of conditions relating to emissions (for more detailed comment, see Miller and Wood 1983; Miller 1993, 1999a; Purdue 1999).

9 The national air-quality strategy, first published in 1997 under Part IV of the Environment Act 1995 and subsequently reviewed, sets objectives for eight pollutants, prescribed in the Air Quality Regulations 1997 (revised 2000). The strategy was necessary in order to comply with (although some of its detailed provisions go beyond) European legislation.

10 Beattie and Longhurst (1999) provide a useful review of progress in urban regions in England, noting that environmental health departments were most likely to take a lead in progressing arrangements within local authorities. Although land-use and transport planners were also involved, the authors suggested that 'insularity of departments' might 'severely hinder' the air-quality management process (*ibid.*: 14). In regional groups (established when problem areas transcend local authority boundaries), land-use planners were found to be less engaged than either environmental health officials or transport planners, a situation that Beattie and Longhurst identify as a significant disadvantage. That, in practice, 'integration' may be more problematic than the rhetoric of sustainability suggests is an important issue, to which we return in Chapter 4.

11 This has proved difficult, at least initially, partly because of conservatism in interpreting the 'new remit', largely because of the uncertainties involved in predicting impacts and, of course, because air quality is never the sole consideration. For example, an appeal against refusal of planning permission for a superstore in Bath on air-quality grounds was upheld in 1996 (for detail, see Miller 1999a).

12 For discussion of planning and renewable energy, especially wind turbines, see Countryside Agency 1999; CPRE 1999b; McKenzie Hedger 1995; National Trust 1999; Royal Commission on Environmental Pollution 2000; and on energy considerations in design and layout, see de Schiller 1999; DETR 1998k; Owens 1986, 1991, 1992; Owens and Cope 1992; Urban Task Force 1999.

13 Regeneration of urban areas has been an important policy objective in its own right, but there are synergies with the efficient use of land.

14 In the 1999 strategy, 'effective protection of the environment' and 'prudent use of natural resources' become part of the economic–social–environmental triumvirate alongside 'high and stable levels of economic growth and employment' and social inclusion (DETR 1999c: para 1.2). While acknowledging the possibility of environmental limits related, for example, to climate change, water resources and fisheries, the strategy points out that 'we cannot protect every bit of the environment for ever' (*ibid.*; para 4.5). This misses a good deal of middle ground, to which we return in Chapter 3.

15 This is not to suggest that social considerations were previously absent – they featured in guidance on development plans (DoE 1992a), housing (DoE 1992b) and retail facilities (DoE 1993b), for example, as well as in the Development Plan Regulations – but that they had become much more prominent as an integral component of sustainable development.

16 Lest they should get carried away, there is a warning for those local authorities who may have 'wider social considerations in mind in considering the future development of their communities'. The 'social' content of the development plan must be limited to 'considerations that are relevant to land use policies' (DETR 1999b: para 4.13).

17 This resentment found particular expression in DoE Circular 22/80 (DoE 1980), which stressed that planning consent should be withheld only when in pursuit of clear purposes and with the economic consequences fully understood.

18 Under the Local Government Act 2000, local authorities have been given a duty to produce a community strategy and a broad power to promote the economic, social and environmental well-being of their area.

19 The statutory purposes of the RDAs, under Section 4 of the Regional Development Agencies Act 1998, are primarily concerned with economic development, employment and regeneration. However, there is a curiously qualified duty (the last in a list of five) for each RDA 'to contribute to the achievement of

sustainable development in the United Kingdom where it is relevant to its area to do so'. At the time of writing, the regional picture remains confusing. It is not clear, for example, which forum will take the lead on regional sustainable development issues, or how planning guidance will in practice relate to the economic strategies produced by the regional development agencies. Up to 1999, only in one English region had the two documents been prepared in parallel (Land Use Consultants 1999). 'Regional Sustainable Development Frameworks', to be prepared following proposals in the government's sustainable development strategy (DETR 1999c), are intended to provide a vision for sustainable development in the region, giving a broad context for regional planning guidance and economic strategies and taking account of the issues raised by them. Regional chambers will contribute to the development of this strategy, adding further complexity. These bodies, established after the 1997 election, are composed of a significant local authority element with representation of non-local authority interests, and are intended to co-ordinate many regional-level activities. RDAs must have regard to the views of the chambers and be willing to give an account of their activities to them.

20 Counsell's study examined the twenty-seven county structure plans in England and Wales that had addressed sustainable development by the time of the study in mid-1995 (58.7 per cent of the total), through documentary analysis, interviews with planners, surveys of participants in the planning process and case studies of examinations in public. He went on to analyse the plans of the 'front runners' separately (Counsell 1999b).

21 Typical of a broad aspiration was the Leicestershire, Leicester and Rutland Draft Structure Plan's 'vision' of minimising the threat to the Earth's climate from increasing emissions such as carbon dioxide and methane (Leicestershire County Council *et al.* 1998). More specific resource-related policies, for example on energy efficiency and water resource management, can also be found in this and many other plans. Quite early on, the Kent Structure Plan sought 'to achieve a sustainable pattern and form of development which will facilitate the conservation of energy and other natural resources, and minimise pollution' (Kent County Council 1993: 14), while the 1998 Newport Draft Unitary Development Plan proposed not to grant permission for development that would have 'a detrimental effect on water quality in watercourses and ground water' (Newport County Borough Council 1998: policy CE11).

22 It is interesting that something like Bedfordshire's policy on land allocation subsequently achieved the sanction of central government, in revised planning policy guidance on housing (DETR 2000b).

23 In an assessment of the effectiveness of the Fifth Environmental Action Programme (CEC 1999c), the European Commission concluded that while the programme achieved some positive results, progress towards sustainable development has been limited. An important reason for this failure was that increasing demand for transport, energy, consumer goods and tourism had frequently outpaced environmental gains from new regulations and efficiency measures. Individual member states recognise the same phenomenon: in the Netherlands, for example, pollution in some areas has been reduced and environmental efficiency improved, but there remain difficulties in meeting objectives in the country's National Environmental Policy Plan for noise, contaminated land, acidification and greenhouse gases, 'mainly because of the consequences which growth in the economy will have for the environment' (Netherlands Ministry of Housing, Spatial Planning and the Environment 1998: 15).

3 Intepreting sustainability

1 Most of our discussion in this chapter relates to human-made and natural capital, since it is these categories that have most obviously been the subject of land-use planning dilemmas. This is not to underestimate the importance of human capital (in terms of knowledge and skills) or social capital (the organisations and structures of civil society): both are of significance in our discussion of the social dimensions of sustainable development.

2 Standard rationales for applying a positive discount rate to future costs and benefits (converting them to (lower) present values) include the fact that capital invested now will be put to productive use; the assumption that society will become wealthier, so that a given sum will confer less utility in future; and the observation that people simply prefer to enjoy benefits sooner rather than later. The validity of the practice has been the subject of intense debate, and dissatisfaction with it was one of the driving forces behind the environmental capital framework (for discussions of discounting in an environmental context, see Goodstein 1995; Pearce *et al.* 1989; and for critiques, see Parfit 1984; J. O'Neill 1993). An important argument for discounting is undermined by removing the assumption of infinite substitutability (Jacobs 1995).

3 In Daly's (1995: 50) view, it makes little sense to talk of substitutability when there is *complementarity* between natural and human-made capital: this means that 'the one in short supply is limiting', so 'we put the constraint on natural capital' (see also Jacobs 1995).

4 The potential for reconciliation depends in part on technological capabilities (Pezzey 1989; Jacobs 1991), including the possibility of substitution of renewable for non-renewable resources (Daly and Cobb 1989; Hartwick 1978), but it also depends on the meaning of development (for example, its content and relationship to material consumption). Important questions, to which we return, are how feasible it is to repair or compensate for losses and degradation (Cowell 1997) and what possibilities might exist for enhancement as environments co-evolve with human development (Norgaard 1994).

5 There has been extensive debate about the social construction of what is 'natural', to which we cannot do justice here (see, for example, Fitzsimmons 1989; Macnaghten and Urry 1998; Smith 1990).

6 Welfare, or well-being, is usually identified in welfare economics with preference satisfaction. Individuals are assumed to have a single, ordered set of preferences and to choose rationally the options that they prefer within their budget constraints – they are utility maximisers. In theory, perfect markets in which such agents interact produce an equilibrium situation where no one could be made better-off without making someone else worse-off – the Pareto optimum, or most efficient, allocation. When markets fail (or are absent), standard cost–benefit analysis, as one of the tools of welfare economics, seeks the same outcome by other means: 'it defines the best policy as that which maximises the "well-being" or "utility" of affected parties' (J. O'Neill 1993: 65).

7 Thresholds often draw upon established concepts of 'critical loads' for pollution or 'maximum sustainable yields' of biological resources, both resonating well with notions of environmental capital.

8 Utilitarianism is the best-known consequentialist ethical theory, according to which the rightness of an act depends on its consequences; in a utilitarian framework we ought to act in ways, and support those institutions, that would result in the maximum amount of (usually human) happiness. While this framework has been much debated (and much criticised) within moral philosophy, it

provides a basis for neo-classical economics and for techniques of modern welfare economics such as cost–benefit analysis (for more detailed discussion, see Hausman and McPherson 1993; Kelman 1990; J. O'Neill 1993, 1998). It also continues to exert powerful influence, not least in planning, and has some stout defenders. Goodin (1995), for example, argues strongly in favour of a form of 'welfare utilitarianism' as a guide to public policy, although his account depends heavily on 'laundering' preferences, an issue to which we return below. Some have argued for non-utilitarian forms of consequentialism: we cannot do justice to these developments here, but see Sen (1987, 1999).

9 More specifically, the various forms of capital are seen as generating flows of benefits, for which the strength of people's preferences can be measured (through revealed and stated preference techniques). It may be accepted (in stronger versions) that some of these benefits are non-substitutable, at least with current knowledge and technology, as discussed above.

10 For some environments – semi-natural habitats are a prominent example – protection cannot be achieved by 'setting them aside and exempting them from use', only by continuing to manage them in an appropriate way.

11 From the perspective of social justice, such limits might also prevent significant harm to minorities being dwarfed by more trivial benefits to the majority (Laslett 1995).

12 Such act-centred ethics seeks to establish certain principles of obligation, or rights, which constrain individual action as well as institutions and practices, and thus 'limits the domains of life in which trading off is … permitted' (O. O'Neill 1997: 131). However, while 'rightness' is not interpreted in terms of maximising the good, this need not mean that it can be characterised *independently* of consequences (Rawls 1972: 30).

13 There are particular conceptual difficulties in providing a sound rationale for environmental protection when interests or rights are construed in an individualistic and anthropocentric way (as they frequently have been). One of the best known is posed by Parfit's (1984) 'non-identity problem': policies adopted now affect not only the state of the environment but also the *identity* of people living in the future; in a sense, therefore, no choice that we make can harm these individuals, because alternative choices would have meant that they did not exist. For a fuller discussion of this and related issues, see O. O'Neill 1996; Norton 1982.

14 See, for example, O. O'Neill (1996), who argues that 'where there is an assumed connection, we cannot exclude from the scope of ethical consideration those others whose shadowy future lives and connection to our activity we acknowledge in acting' (*ibid.*: 116). Furthermore, 'activity that will be imprinted on the enduring natural and social world will be predicated upon quite powerful assumptions about its remote as well as its immediate recipients' (*ibid.*: 118).

15 This account draws upon the Kantian requirement that action and reason follow principles that are thought of as adoptable by all (that is, principles that are universalisable; see O. O'Neill 1996, especially chapter 2.3). Since a principle of injury self-evidently does not meet this requirement, its rejection is a matter of obligation. On this account, justice *requires* a certain amount of environmental protection, and this is a matter of perfect obligation, matched by rights; 'its demands fall on all and are owed to all' (O. O'Neill 1996: 184). In much modern liberal thought, rights (which may include environmental rights; see Beatley 1989) are seen as the fundamental ethical category, with duties following as a counterpart. However, Onora O'Neill (1996, 1997) argues that at least as much can be arrived at by starting with obligations (for an alternative perspective, see Sen 1999). She contends further that a duty-based approach is

more promising in relation to dispersed or abstract features of the natural world (see, especially, O. O'Neill 1997).

16 Individual non-human animals might also be beneficiaries, but within this Kantian framework they can be seen as the subject of indirect duties arising from the direct duty of human agents to cultivate a good disposition in themselves (for a more detailed exposition of this rather complex argument, see O. O'Neill 1998).

17 Interestingly, in the UK, the Labour government elected in 1997 promised 'a right to clean air' but later argued, in terms of costs and benefits, that certain demanding air-quality targets could not be justified (ENDS 1999c). The US Clean Air Act limited the application of cost–benefit analysis in the setting of air-quality standards (for a discussion, see Sagoff 1988).

18 For a full discussion, see O. O'Neill (1996), especially chapter 7.2. Principles underlying a range of required social virtues are derived from the premise that agents have reasons to reject not only the principle of injury (a matter of justice) but also the principles of indifference or neglect. It is insufficient not to damage environments, because unless they are also understood, cared for and cherished, human vulnerabilities will multiply. Environmental virtues may take different, context-dependent forms including, for example, preservation of cultural landscapes and protection of uncultivated wilderness or endangered species.

19 In one sense, all ethical *reasoning* is anthropocentric, but critics of 'anthropocentrism' are often troubled by what might more accurately be called 'speciesism' (O. O'Neill 1997), in which humans are accorded special ethical status, and/or by instrumentalism in thinking about, and treatment of, the non-human world (Hargrove 1992).

20 This argument is developed within a broadly Aristotelian tradition, in which 'the purpose of politics is the good life of its citizens' (J. O'Neill 1998: 16–17); for Aristotle, a good life was characterised in terms of the virtues.

21 As the foregoing discussion has shown, the distinction is not only between 'those who insist on the need to establish and maintain a minimum environmental stock or capacity and those who accept the inevitability of trade-offs' (House of Lords Select Committee on Sustainable Development 1995: 10); there are important differences within the former group on *why* such a stock should be maintained and (therefore) on what constitutes the 'minimum' – that is, what should be counted as critical.

22 As expressed in the sentiments of protesters against a proposed opencast coal mine in Nottinghamshire, who objected to the promised restoration on the grounds that British Coal's 'designer landscapes' could not 'compete with Nature' (Donnison 1990: 76).

23 Although such considerations were implicit, at least to some extent, in both economic and environmental domains. Basic environmental protection, for example, tends to be progressive because the poor are typically exposed to the most degraded environments. The link was more explicit in Brundtland's neo-material rationale for environmental protection: many of the world's poorest people depend directly on natural resources for subsistence (see also Pearce 1988). However, the argument does not transfer straightforwardly to less material and non-instrumental concerns.

24 We do not intend to suggest that planning authorities would always adopt environment-led approaches, but the example is not implausible (as illustrated in Chapters 2,3 and 7).

25 It is widely accepted that the satisfaction of purely want-regarding preferences – seeking to maximise pleasure – does not necessarily promote the welfare of the individual. Many theorists have therefore substituted prudent or informed pref-

erences, or 'welfare interests' (Goodin 1995: 20) for pleasure or hedonistic desires. But to accept that preferences need to be 'laundered' (Hausman and McPherson 1993: 714) suggests that they can be judged by criteria that are independent of the preferences themselves. For John O'Neill (1998: 48), this is no more than 'a disguised way of talking about objective determination [of the good]'.

26 In his influential book *The Economy of the Earth*, Mark Sagoff (1988) argues that public policy should not be based on the aggregation of people's preferences as 'consumers' but informed by the inter-subjective judgements they make when deliberating as citizens. John O'Neill prefers to talk of 'ideal regarding preferences' emerging out of deliberation. While their respective critiques of cost–benefit analysis share much common ground, O'Neill (1993: 175) feels that Sagoff fails to provide a convincing account of the boundary between politics and the market, given that 'there are very few economic decisions that do not have environmental consequences'. O'Neill is more fundamentally critical of the market as an institution (see, especially, J. O'Neill 1998).

27 There is an immense literature on this subject, which we cannot go into here. While some retain an aversion to distinguishing between wants and needs, many have perceived a morally significant difference between 'goods of the needs category' and 'goods of the wants category' (Wissenburg 1998: 207; for a useful discussion, see Hausman and McPherson 1993).

28 We do not overlook the fact that rationales can be deployed strategically, or even cynically. Just as environmentalists have been ready to employ the instrumental, utilitarian arguments that many of them reject in principle, countless developments have been defended on the grounds that they will promote the 'greater good', or social justice, when in actuality they serve narrow sectional interests. On the one hand, this demonstrates that opponents feel a need to legitimise their claims (sheer 'powering' will not do), and it opens up the various arguments to scrutiny and debate. On the other hand, later chapters in this book suggest that it is conservationists who have had to search longest and hardest for rational justification of their case (hence the explosion of interest in environmental ethics), while the merits of 'development' have often been taken for granted. Recourse to rationality, as Flyvbjerg (1998) suggests, may be inversely proportional to one's possession of power.

29 When demonstrably important claims conflict, the question might turn on whether particular social needs could be met by alternative technical, spatial or institutional solutions, or even in quite different and novel ways. Where conflict is unavoidable, 'the demands of principles that have been flouted or skimped may be acknowledged (although they often are not) by responses that range from apology or confession, by way of restitution or reparation, to regret or remorse' (O. O'Neill 1996: 160). There are clearly interesting implications here for the concept of environmental compensation.

30 Whereas classical liberalism espoused an objective concept of the good, modern liberalism tends to reject objectivism (beyond the imperatives of liberty and autonomy) as elitist or paternalistic (for a useful discussion, see J. O'Neill 1998). Meta-ethical sceptics – among whom we can include full-blooded postmodernists and many liberals of the Austrian School – deny the possibility of any rational justification for ethical judgements (Keat 1997: 44).

31 Echoing Dr Pangloss, in Voltaire's *Candide* (1759, chapter 1): 'Dans ce meilleur des mondes possibles ... tout est au mieux' (all is for the best in the best of all possible worlds).

32 This is not just due to liberal squeamishness about making judgements of value but reflects powerful presumptions about property rights embedded in planning

law, despite the fact that the 1947 Town and Country Planning Act effectively nationalised the rights of development in land.

33 In policy guidance (DoE 1996c), local planning authorities are advised to consider the need for further retail and leisure facilities in their areas and must then apply a sequential test in finding sites, with town centres being preferred and sites further out being considered in descending order. Where there is not an up-to-date development plan, developers bringing forward a proposal must demonstrate 'need' and show that they have applied the sequential approach. Critics argue that this is all too difficult, suspect the government of protectionism (towards operators in town centres) and accuse it of elitism for adopting the values of 'those who neither go to shopping centres nor enjoy commercial leisure activities' (Lock 1999: 111).

34 "'When *I* use a word", Humpty Dumpty said [to Alice], in a rather scornful tone, "it means just what I choose it to mean – neither more nor less"' (from Lewis Carroll's *Through the Looking Glass and What Alice Found There*).

4 Defining and defending: approaches to planning for sustainability

1 A chapter in *Agenda 21* (UNCED 1992a) identifying the role of local government in a transition towards sustainable development. By 1999, about a third of the local authorities in England, Scotland and Wales had produced a Local Agenda 21 strategy, and a further 45 per cent intended to produce one by the end of 2000 (Farmer *et al.* 1999).

2 This is particularly important where a requirement for environmental assessment opens up previously closed decision-making processes. In the UK, for example, assessment has been extended to certain categories of development (such as offshore oil and gas installations) that were not covered by the town and country planning legislation.

3 It is expected that the Directive will be adopted by spring 2001, and that it will leave considerable discretion to member states.

4 For example, the Commission has undertaken to develop methods for SEA (together with socio-economic assessment) in evaluating trans-European transport networks (Bina 2000). Some member states, for example the Netherlands and Denmark, apply an 'environmental test' to draft legislation. 'Strategic Environmental Impact Assessment' is a statutory requirement in the Netherlands for sectoral and spatial plans that determine the location of projects for which an environmental assessment is already required. For further discussion of SEA in a number of countries, see DETR (1998e), and for an overview of European member states' SEA practices in the transport sector, see Bina and Vingoe (2000).

5 It is interesting to note, for example, that when the decision was made in November 2000 to reduce excise duties on ultra-low sulphur fuels (see Chapter 5) and presented as being environmentally motivated, the Treasury did not calculate the effects of its proposals on carbon dioxide emissions (ENDS 2000c).

6 Interestingly, they have sometimes rejected the capacity approach because it might *fail* to provide solid support for such arguments. In one English county, planners felt that it 'would not provide the numerical answer to the housing capacity debate people were looking for ... that the county was "full"' (Counsell 1999b: 50, 55).

7 One might argue that the government fulfilled its proper role in adjudicating between different needs, choosing on this occasion to give priority to a social

need for housing. But this begs important questions about 'need', particularly in the context of intense market-driven demands for development in the south-east of England, another 'black box' issue.

8 Connell (1999: 8) notes that the declared aim of some planning officers was to 'expose and make explicit' the consequences of development choices, especially those with serious and irreversible effects.

9 Methods for deriving monetary valuations of environmental quality feature prominently in national guidance on policy appraisal (UK Government 1990; DoE 1991, 1994e) and have become important considerations in environmental regulation, notably in the duty on the Environment Agency to 'take into account the likely costs and benefits' of exercising its powers (Section 39(1) of the Environment Act 1995). However, guidance supporting the Act acknowledges that many costs and benefits are inherently difficult to quantify in monetary terms, and that judgements are still required in such analyses (see discussion in Jewell and Steele 1996).

10 A good example is the tax on primary aggregates, discussed in Chapter 7.

11 According to the DoE (1994e: para 3.22), analysis of the economic and environmental costs of forestry expansion 'contributed significantly' to the reorientation of government forestry policy towards multiple-use benefits.

12 The statutory agencies have nevertheless thought it politic to keep abreast of, or even sponsor, developments in environmental valuation techniques (see, for example, Burney 2000). They have usually preferred to use them in conjunction with *a priori* constraints, for example in insisting that some habitats are irreplaceable and should be removed from the arena of trade-off, while emphasising the role of techniques in estimating replacement costs for others – in effect, in costing a 'no net loss' policy (for further detail, see Shepherd and Harley 1999).

13 Indeed, some of the most powerful political support for cost–benefit analysis has come from governments keen to reduce the costs of environmental regulation (Jewell and Steele 1996; Yeager 1991).

14 Most land-use planning systems make legal provision for public consultation and have been seen as something of an exemplar in this respect, at least in comparison with other aspects of environmental planning (TCPA 1999).

15 The opportunity to be heard, at least, has become enshrined in declarations of human rights. For example, Article 6 of the European Human Rights Convention, incorporated into the Human Rights Act 1998 in the UK, states that 'in the determination of his civil rights … everyone is entitled to a fair and public hearing within a reasonable time by an independent and impartial tribunal established by law'. The provisions of the Aarhus Convention on Access to Environmental Justice are also relevant, although this convention is not yet in force at the time of writing.

16 One investigation revealed two deep-rooted professional stances towards public involvement: it was either conceived broadly and considered integral to planning, 'demanding open-ended approaches and generating added value', or seen as an optional, technical input 'to a process controlled by professionals and/or developers' (DoE 1994f: para 9). Critics tend to see predominance of the latter stance as having contributed towards failure.

17 Because of the adversarial and often intimidating nature of procedures, the unrealistic time and resource demands on many would-be participants, and the predominance of professional expertise and language. Examination in public of development plans is subject to some, but not all, of the same problems.

18 While formal linkages have not always been well defined, local authority planning departments have sometimes played a key catalysing role in Local Agenda

21 ventures, especially where these involve exercises such as 'visioning' and 'planning for real'.

19 For example, continuing to hold meetings in relatively inaccessible locations (council offices in the evenings) and controlling dialogue through 'mostly male, mostly "expert" gatekeepers' (Buckingham-Hatfield 1997: 215). Other studies seem to confirm that 'new sustainability initiatives largely draw upon the "old faces"', even if they attract small numbers of genuine newcomers (Selman 2001: 27).

20 Following the view that particular institutions (including markets) *create* particular forms of value (see, for example, Blake 1999; Hausman and McPherson 1993; J. O'Neill 1995; Rawls 1972). The interesting question is how malleable values are when individuals move between different institutional contexts.

21 Areas were selected either on the grounds that they already suffered serious environmental problems or because they had important environmental qualities deemed worthy of protection.

22 Conventional measures of 'development' still dominate. Cole (2000) notes, for example, that six out of eight English regional development agencies had adopted targets for increasing GDP, with some seeking higher rates than the expected EU average.

23 Degeling's specific concern was with urban public health agendas in Australia.

5 Moving targets: planning for an integrated transport policy

1 Although there was still a strong emphasis on the completion of trans-European road networks.

2 A phrase famously attributed to Margaret Thatcher.

3 As it was characterised by the then Transport Minister, Paul Channon, at the 1989 Conservative Party Conference.

4 There had been longstanding critiques of traffic forecasting and the roads programme, but these had made little impact on the prevailing 'predict and provide' mentality (Adams 1981; Hamer 1987; Tyme 1978).

5 SACTRA's origins lie in the Leitch Committee, set up by the Secretary of State for Transport in 1976 to examine the Department's method of appraising trunk road schemes and later effectively reconstituted as a standing committee. Dudley and Richardson (1996a: 75) suggest that the 1976 committee, whose membership included the Director of the Civic Trust, 'represented one of the few examples of the environmental lobby penetrating the core network' and that it provided 'the one example in the past fifty years' of the kind of prestigious, apolitical forum that might facilitate policy-oriented learning across belief systems (Sabatier 1987). We would add the further example of the Royal Commission on Environmental Pollution (for a discussion of the role of this body in general, see Owens and Rayner 1999).

6 In his theory of how public policies develop and change, Kingdon argued that process streams of problems, policies and politics flow largely independently through the political system.

7 Foreword by John Prescott, Deputy Prime Minister and Secretary of State for Environment, Transport and the Regions.

8 This is not to suggest that transport planning in preceding decades had been uniformly in favour of road-based mobility. The 1968 Transport Act included measures to support rail and created passenger transport authorities to co-ordinate public transport in conurbations (Truelove 1999). But the main *thrust* of

policy was to accommodate traffic growth, and other measures had little real impact on the decline of public transport. Non-motorised transport received hardly any attention at all. A number of other European countries (including France, Germany and the Netherlands) were more supportive of public (and in some cases, non-motorised) transport (Simpson 1987), but they too failed to curb the massive growth in car ownership and use.

9 Introduced by the Conservative government at a rate of 5 per cent per annum in 1993, increased to 6 per cent by the Labour government soon after taking office in 1997.

10 Significantly, the academic most closely associated with the concept of 'new realism' (Professor Phil Goodwin) chaired a small group of expert advisers to ministers during preparation of the White Paper.

11 Sabatier (1987) suggests that in any policy sub-system, there is an enduring 'policy core'. Learning induced by argument, evidence and persuasion may not be sufficient to bring about change to the core; this is more likely to result from significant developments external to the sub-system.

12 Still (1996) provides an interesting account of the views of land-use and transport planners. The former were generally more sceptical about the benefits of modelling than the latter, and UK land-use planners were less enthusiastic than their US counterparts. Some of those interviewed felt that modelling perpetuated the *status quo*, others that 'numbers' were a vital element of any scheme justification (an interesting example of the 'analytical arms race' that we discussed in the previous chapter). A widely held, if somewhat inconsistent, view was that 'models were good for educative purposes and informing, but bad if you believe the answers' (*ibid.*: 567). Richardson and Haywood (1996) offer a useful discussion of models of the impact of different road and rail investment strategies for a trans-Pennine route in the UK. The authors criticise the narrow framing of the problem and the assumptions underlying the models used, but they also observe that 'the final decisions would always be politically determined, perhaps validated but never overturned by an experimental transport modelling exercise' (*ibid.*: 44).

13 Particularly with growing car ownership and use and increasing movement of freight by lorry.

14 Significant factors include increasing female participation in the labour force, new patterns of consumption and the introduction of a market ethos into the provision of public services (for example, increasing emphasis on 'parental choice' in education).

15 Whereas Sabatier (1987) envisaged policy learning taking place through the interaction of advocacy coalitions, whose members share basic values and beliefs, Hajer (1995) envisaged looser discourse coalitions, mobilising around a story-line, which might successfully challenge dominant discourses to bring about policy learning and change.

16 The Department of Trade and Industry (DTI 2000) anticipates a 28 per cent increase in energy consumption by road transport and an 18 per cent growth in carbon dioxide emissions between 1996 and 2010 (central economic growth and high energy prices scenario). The Commission for Integrated Transport (CfIT 1999: para 11), drawing on work by the DETR, thought it possible to return to 1996 levels of carbon dioxide emissions by 2010 if voluntary agreements with vehicle manufacturers were delivered, or even to achieve a modest reduction, but only with 'intensive' application of the policies in the Integrated Transport White Paper, which seems unlikely. The Royal Commission on Environmental Pollution was sceptical that such reductions could be achieved (RCEP 2000).

17 In part based on expert advice that contributions from mainland Europe were more significant than had previously been thought. It was claimed that it would not be feasible to meet the objective within the time frame of the strategy 'even if all motor vehicles were taken off the roads' (DETR 2000a: para 262), but a new objective is to be given 'high priority' (*ibid.*: para 270).

18 In terms of our discussion in Chapter 3, we could say that vehicle emissions standards draw the 'criticality boundary' imperfectly because they are not based on the capacity of the receiving environment.

19 This perspective gained ground after SACTRA (1994) confirmed the traffic-generating propensities of new roads, and it clearly influenced the 1998 Transport White Paper, in which major new road construction was presented as a last resort. Even the tenacious belief that new infrastructure stimulates local or regional economies was called into question: 'roads operate in two directions, and in some circumstances the benefits will accrue to other, competing regions' (SACTRA 1999: 22). SACTRA (*ibid.*: 17) maintained that 'generalisations about the effects of transport on the economy are subject to strong dependence on specific local circumstances and conditions'. A similar point had been made by the Leitch Committee more than twenty years earlier (DoT 1977; Truelove 1999). Policy learning can indeed be a slow process.

20 In the UK, the change of attitude over twenty years or so was striking. At a press conference to launch a consultation paper in 1976, the Transport Minister responded to a query about land use by suggesting that it was 'hardly a question you can expect me to answer' (quoted in Smith, J. 1994: 91). Still (1996: 566), in a review of practice, quotes one planner as saying 'when I came here in 1975, the view of things was that transport was quite separate, and of no consequence to land use planning'.

21 The focus here is on roads, which dominated infrastructure planning in the period covered, but developments such as railways, ports and airports raise many similar issues.

22 Given that restraining traffic in a more general sense was not on the agenda, Buchanan's 'solution' was large-scale investment in infrastructure to separate traffic from other aspects of urban activity. The reality in many urban areas was the 'middle course' that Buchanan had feared: 'poor traffic access and a grievously eroded environment' (Cullingworth and Nadin 1994: 229).

23 With a few exceptions. Glasgow, for example, continued to build urban motor-ways, with central government assistance for economic regeneration (Truelove 1999).

24 One of the ironies of the appraisal system was that land of low economic value was attractive to road planners, although it might be of ecological or cultural significance (RCEP 1994).

25 Areas of intensive farmland were 'recreated' as downland meadow and wetland, involving complex operations to denutrify the soil and replant with appropriate species. As we have argued in Chapter 3, such 'compensation' cannot necessarily replace what is lost; nor does it vindicate a decision to proceed with the development. It is, perhaps, 'the best thing to do in the circumstances' (Don 1999: 50).

26 A proposed East London river crossing, the approach to which would have destroyed part of the ancient Oxleas Wood, for example, was not pursued.

27 The economic and environmental analyses remained separate, in acknowledgement of the many difficulties associated with attributing monetary values to environmental assets.

28 The agencies' view is that many environmental services are not substitutable on a meaningful timescale, so that targets and limits 'are a key input to the environmental capital methodology' (CAG and Land Use Consulatants 1998: 46).

29 Decisions on 14 of 147 inherited schemes had been made in an accelerated review in 1997 (three, including the Salisbury by-pass, were dropped).

30 Some of the schemes withdrawn were on routes for which responsibility was being transferred to local level, leaving it to local highways authorities 'to decide whether to take them forward as local road schemes' (DETR 1998g: 106). Seven schemes from the inherited programme were to be 'progressed pending final decisions', and forty-four were 'to be considered under the new approach to appraisal by regional planning conferences as transport problems requiring broad solutions, not necessarily road solutions' (DETR 1998g: 102). A number of acknowledged transport problems were to be addressed in a series of proposed studies also announced in the roads review: some would involve localised consideration of problems on particular routes, but others would be large-scale 'multi-modal' studies led by government regional offices in consultation with regional planning conferences. It is interesting to note that the first of the multi-modal studies to reach the stage of a preliminary decision – the 'Access to Hastings' study – did after all result in a recommendation for road building. The South East Regional Assembly voted in February 2001 to advise the Secretary of State to proceed with the controversial eastern and western by-passes around Hastings as part of a package of transport measures to improve access and (it was argued) enhance prospects for local regeneration.

31 It is also significant that in the Local Transport Settlement 2000 (through which central government allocates resources for transport to local authorities), eighteen of twenty-one major schemes that won funding were solely roads projects (*Surveyor* 2000).

32 Often in the face of opposition from commercial interests. However, a typical experience in central areas has been for support to grow once schemes are implemented.

33 Among other measures, developments were to be concentrated in locations well served by public transport and within 400 metres of a bus stop or regular public transport service, and development control policies would employ maximum standards for parking.

34 Many planners have thought it crucial that where public transport does not already exist, its provision should be integrated with the development of land (for an interesting case, see Municipality of Schiedam 1991). This point seems to have been lost in revised planning guidance, which talks of relating new development to public transport or *potential* public transport (DETR 1999f).

35 These supplemented design guidelines (DoE and DoT 1992, first published 1977), which had come to be seen as overly car-oriented in favouring *cul-de-sac* layouts.

36 Sometimes to the exclusion of other dimensions of sustainability. Green field (or even green belt) development, for example, has come to be legitimised on the grounds that settlements would be 'sustainable' – that is, planned in ways that might reduce the need to travel, or located in rail corridors. For a critique, see Miles 2000.

37 Those with populations between 10,000 and 100,000 were variously reported to be efficient, with some emphasis on the 20,000–30,000 range (Owens 1986; Owens and Cope 1992).

38 Similar principles may be found elsewhere in Europe, for example in development of the Copenhagen 'Fingerplan' in the 1980s and 1990s.

39 'A' locations were the most accessible by public transport (usually in city centres) and 'C' locations the least.

40 'Making the Land Use, Transportation, Air Quality Connection' (LUTRAQ).

41 In another interesting example of transnational policy learning, the then attorney for 1000 Friends of Oregon and key figure in the LUTRAQ Project was involved in research and policy networks spanning European experience.

42 Including the M3 extension at Twyford Down in Hampshire, the Newbury by-pass in Berkshire and the Birmingham northern relief road.

43 A further example of 'learning'. LUTRAQ received national awards for planning from the American Planning Association and the US Environmental Protection Agency. Those associated with the project have made many presentations throughout the USA.

44 For one striking example, when the Dutch Environment Ministry moved to an 'A' location adjacent to the Central Station in the Hague, the percentage of employees travelling to work by car fell from 41 to 4 per cent and the proportion using public transport increased from 34 to 77 per cent (M. Post, Netherlands Ministry of Housing, Spatial Planning and the Environment, personal communication 1994). However, the ministry's adoption, at the same time, of restrictive parking policies must have strongly reinforced the locational factor.

45 The figures were 37 per cent for 'C' locations, 16 per cent for 'B' locations and only 1 per cent for 'A' locations. Moreover, the number of business areas in 'residual' locations grew faster than those in the classified 'ABC' locations, indicating that much employment growth was taking place outside urban areas.

46 ETRAC cites the Department of Trade and Industry's promotion of high-technology 'clusters' on green field sites.

47 The reaction in Sheffield (Kay 1999: 1). Reasons for resistance included the fragility of the city's economy – 'any attempt to restrict access was likely to wreck its recovery' – and a perception that the government 'was dodging its responsibility by passing the handling of the controversy to local authorities'. By the end of 1999, twenty-seven councils had expressed an interest in congestion charging or parking taxes (ENDS 1999a) but one survey suggested that many larger metropolitan authorities would be deeply reluctant to adopt such measures in advance of substantial improvements in public transport provision (Radio 4 *Today* programme, 1 November 1999). Central government could withhold resources from authorities whose transport strategies are not in line with integrated thinking, although this would seem to be at odds with the notion of local self-determination. It will also be necessary to agree on regional frameworks if local authorities are not to try to outdo each other with motorist-friendly policies.

48 The Road Traffic Reduction Act 1997, the result of a Private Member's Bill, was supported by the government after its sponsors agreed *inter alia* to delete any reference to national traffic-reduction targets. It requires local authorities to draw up plans for reduction in road traffic or its rate of growth unless they can demonstrate good reason for not doing so.

49 Crenson (1971) used the concepts of 'un-politics' and 'non-decision making' in his classic study of corporate power and air-pollution policies in two locations in the United States – Gary, Indiana, and East Chicago.

50 Fuel duties were increased only in line with inflation, and vehicle excise duty for some lorries was cut, even though it had been raised only in the previous budget to reflect the damage caused to roads (ENDS 2000a). Environmental measures were modest, longer term or less visible (including progress on graduating vehicle excise duty for cars and reforms to the much-criticised company car taxation regime). An additional £280 million was made available for transport, £28 million of which was promised for new road schemes. There was also comfort for the haulage industry in the government's decision to authorise 44-

tonne lorries from January 2001, a move long resisted by many environmentalists.

51 From which all modes would benefit, especially rail. But the plan envisaged substantial spending on roads (£60 billion), including a hundred new by-passes and 360 miles of trunk road and motorway widening, enough for some critics to accuse the government of 'a return to "predict and provide"' (ENDS 2000b: 19).

52 In the case of petrol, this was conditional on the oil companies guaranteeing nationwide access and would arguably engineer a shift to cleaner fuel. For diesel, the measure represented a straightforward reduction in duty, since ultra-low sulphur diesel was already in effect the only form available in the UK.

53 Directly, in the case of the hauliers and farmers in September 2000, but arguably they were supported by the economic power of the major oil multinationals, whose apparent timidity in defying the pickets led to accusations of collusion, vigorously denied by the companies (Hetherington *et al.* 2000; Macalister and Elliott 2000; Maguire 2000).

54 It might also be argued that the new realism itself, and to a considerable extent the White Paper, have a fundamentally urban bias, paying relatively little attention to the more structurally difficult issues of rural transport (for example, the absence, in many places, of adequate – or even any – local facilities). For discussion of some of the issues, see Cullinane and Stokes 1998; Farrington *et al.* 1998.

55 Set up to advise on integrated transport policy in 1999, in fulfilment of a commitment in the White Paper.

56 Although even then, we might note, this might not be reflected in GDP as the standard measure of welfare.

57 Or they might question the way in which welfare is characterised (for example, that it is anthropocentric). In practice, what often happens is that opponents argue interminably over the nature and measurement of the costs and benefits. Such disputes may be a proxy (an inferior one, according to Sagoff (1988)) for more fundamental debate.

58 The claim that preferences reflect 'consumer choice' (Breheny 1995: 92) is itself a substantial one. Preferences have often been reinforced or stimulated by public policies that in turn reflect powerful corporate interests. Investment criteria favouring roads, a generous taxation regime for company cars and permissive land-use policies could be said to have *shaped* preferences and not just responded to them. Still, this may make the preferences no less potent.

59 Following Isaiah Berlin's distinction in his essay, 'Two concepts of liberty'. Negative freedom, he argues, concerns the sphere within which a subject 'is or should be left to do or be what he is able to do or be, without interference by other persons'. Positive freedom is involved in the answer to the question 'What or who is the source of control or interference that can determine someone to do, or be, this rather than that?' (Berlin 1969: 121–2).

6 Planning for biodiversity: ethics, policies and practice

1 In Britain, statutory arrangements for nature conservation encompass the protection of geological features. Because of our emphasis on biodiversity, we do not refer explicitly to such features in this chapter, although some of our arguments would apply to them.

2 Defined in the 1992 Convention on Biological Diversity as:

The variability among living organisms from all sources including, *inter alia*, terrestrial, marine and other aquatic ecosystems and the ecological complexes of which they are part; this includes diversity within species, between species and of ecosystems.

UNCED 1992b

3 'Nature', as Raymond Williams (1976: 184) remarks, 'is perhaps the most complex word in the language'. We have sympathy with David Wiggins' (1999: 10) view that 'Nature may be understood not as that which is free of all trace of our intervention – in England few things could pass such a test – but as that which has not been entirely instrumentalised by human artifice'.

4 English Nature, the Countryside Council for Wales, Scottish Natural Heritage and the Environment and Heritage Service in Northern Ireland (advised by the Committee for Nature Conservation and the Countryside). Divergence between practices and policies in the constituent parts of the UK may be increased by devolution of responsibilities for nature conservation to the Scottish Parliament and the National Assembly for Wales in 1999 and to the Northern Ireland Assembly in 2000.

5 EC Council Directive on the Conservation of Natural Habitats and of Wild Fauna and Flora (Directive 92/43/EEC).

6 EC Council Directive on the Conservation of Wild Birds (Directive 79/409/EEC).

7 Governments in the European Union have been required to consult on possible SACs before submitting lists to the European Commission, at which stage the sites become candidate SACs. Once agreed by the Commission they become Sites of Community Importance, with full protection under the Directive. Under the legally binding timetable set out in the Directive, the process of formal designation of SACs should be complete by June 2004. When considering potentially damaging land-use change, candidate SACs and potential (not yet classified) SPAs have to be treated as if they were already designated as European sites.

8 All national nature reserves, Ramsar sites (wetlands designated under the 1971 Ramsar Convention, amended 1982), SPAs and SACs are (or are likely to become) SSSIs under national legislation.

9 Some local nature reserves are also SSSIs, but the latter are usually sites of at least national significance.

10 English Nature (1997b) has claimed that the biggest single cause of degradation or loss is overgrazing in the uplands as a result of financial incentives paid under the Common Agricultural Policy.

11 For example, Nature Conservation Orders (Section 29) and Limestone Pavement Orders (Section 34).

12 Certain land-use changes normally covered by the Permitted Development Orders, but which require assessment under the Environmental Assessment Regulations, can no longer proceed without a full planning application being made. The Habitats Regulations 1994 also impose restrictions on 'permitted development' likely to have a significant effect on a European site. However, the UK government has been criticised for failing to regulate conversion of uncultivated and semi-natural areas to intensive agricultural use, as required by European Environmental Impact Assessment Directives 85/337/EEC and 97/11/EC (CPRE 2000a; English Nature 2000).

13 The Hedgerow Regulations 1997 afford some degree of control over removal of farmland hedges of ecological, historical or cultural significance. The local planning authority has to be notified of intended removal of a hedgerow, has six weeks in which to assess its importance against eight criteria (at least one of which must apply) and if necessary can issue a retention notice. Critics argue

that the regulations leave a high proportion of hedgerows unprotected and fail to take into account their landscape importance.

14 Schemes offering annual payments for positive management include the Environmentally Sensitive Areas Programme in England, Scotland and Northern Ireland, Tir Gofal (in Wales) and the Countryside Stewardship scheme in England (for details and further examples, see Farmer *et al.* 1999). The spatial extent of such schemes is not great, however: in England and Wales, they cover 7 per cent of agricultural land (Environment Agency 2000).

15 However, Environmental Impact Assessment and Habitats (Extraction of Minerals by Marine Dredging) Regulations are being prepared by the DETR at the time of writing. Some forms of renewable energy are likely to raise similar issues (RCEP 2000).

16 Although the agencies prefer to present such exercises as 'objective' description and mapping, they clearly embody particular values (in this case that it is right to protect 'character' or biodiversity) and seek to incorporate such values into the decision-making process.

17 Enhancement may also involve social equity considerations when there are deficits in access to areas of wild space.

18 Estuary strategies are advisory documents, sponsored and produced by a wide variety of stakeholders, which attempt to integrate economic and environmental concerns in estuarine environments.

19 English Nature's general definitions were as follows:

- *Critical natural capital* is defined as those assets, stock levels or quality levels that are highly valued; and also either essential to human health, essential to the efficient functioning of life support systems, or irreplaceable or unsubstitutable for all practical purposes.
- *Constant natural assets* are those aspects of the natural capital resource which are not critical in themselves but contribute to the overall quality of our natural capital which should not decline below defined levels. It is the safeguarded total levels that should be maintained and not necessarily all of the individual component features.

Shepherd and Gillespie 1996: 7

20 Required social virtues may demand even greater protection (see Chapter 3).

21 Whether intrinsic values are objectively there in the natural world, regardless of the presence of human beings to recognise or respect them, has been a matter of intense but inconclusive debate. Onora O'Neill (1997: 127) considers it too metaphysically demanding to show that such values exist as 'part of the furniture of the universe'. This difficulty is partly circumvented by characterising intrinsic value as (unavoidably) human-centred, but non-instrumental. Many authors accept that we value non-human entities because we recognise that they have goods of their own and believe that (in most cases) such naturalness is good: Hargrove (1992: 141) calls this 'weak anthropocentric intrinsic value' (see also Goodin 1992; J. O'Neill 1993).

22 Some of the best-known arguments for extending ethical standing to nonhumans employ more exacting criteria, such as self-awareness or sentience, tending, as Johnson (1991: 197) observes, to incorporate beings with at least some 'human-like qualities'. Those following Peter Singer (1976, 1979) would take account of the pains and pleasures of sentient animals within a utilitarian framework; others campaign tirelessly (and with some degree of success) for the attribution of certain legally enforceable rights to at least some non-humans, particularly the higher primates (for a detailed exposition, see Wise 2000).

Others, like Johnson (1991), have sought to demonstrate that virtually all living things (including collective entities) have identifiable 'goods of their own' in terms of conditions that are constitutive of, or conducive to, their flourishing, and (furthermore) that having such interests is sufficient to endow them with (direct) moral considerability (Attfield 1987; Elliot 1993; Goodin 1992; Goodpaster 1978; Hargrove 1992; Johnson 1991; Taylor 1986). Such principles still leave us to deal with the problems of 'relative moral significance, distribution, and procedural moral rules' (Johnson 1991: 198–9; Regan 1981), extending into a new realm the more familiar difficulties encountered when different human interests or rights conflict.

23 This perspective might help with non-sentient or dispersed features of the natural world too: if people's lives are enriched by their ability to appreciate value in nature, we should value even those features that could not be said to have 'goods of their own', or interests. Johnson (1991) cites Ayer's Rock and the Grand Canyon as examples of such features.

24 Schemes can only be considered on these grounds further to an opinion from the European Commission, following consultations by the government of the member state concerned.

25 As we have argued in Chapter 3, to reject utility-maximising principles as the sole basis for interpreting sustainable development is not to insist on certain courses of action *whatever* the consequences. In the context of nature conservation, it is generally acknowledged that nothing in a democratic society can be deemed inviolable in all circumstances (Pritchard 1994). Proponents of 'safe minimum standards', on which targets in the UK *Biodiversity Action Plan* have been based, allowed that they might be transgressed in (unspecified) circumstances where the costs of protection were 'unacceptably large' (Bishop 1978: 10).

26 Conservation bodies lobbied for the strongest presumption against damaging development in changes to legislation (see, for example, English Nature 1998c). The Countryside and Rights of Way Act 2000 imposes a statutory duty on local planning authorities to further the conservation and enhancement of SSSIs, and it modifies development control procedures in ways that are likely to require secondary legislation and 'radical rewriting' of the relevant planning guidance (RSPB 2000: 1).

27 An audit was conducted by a partnership, brought together by Peterborough Environment City Trust, which also included Peterborough Wildlife Trust and English Nature. Existing criteria for designated areas were the major factors informing what was considered 'critical' and identifying targets, with public enjoyment relevant to the identification of 'constant natural assets'. However, critical status was extended to county wildlife sites.

28 This covered thirty key habitats and thirty-nine endemic or globally threatened species and was conducted jointly with the South West Region Biodiversity Partnership; English Nature and the Environment Agency also joined in the production of the action plan.

29 Although the overall extent to which local nature reserve powers have been taken up has been 'patchy and limited' (English Nature 1998d), there was a substantial increase in activity after the mid-1980s (UK MAB Committee 1998). As of 1999, there were over 620 statutory LNRs (RTPI 1999). Of those reserves declared in 1991–95, 80 per cent were in urban (30 per cent) or urban fringe (50 per cent) areas (UK MAB Committee 1998). It has been proposed that not less than one hectare of LNR per thousand population should constitute a minimum standard of provision for accessible green space (Box and Harrison 1993; English Nature 1996).

30 Although not far enough to make nature conservationists complacent (see, for example, English Nature 1999b).
31 Case C-57/89 Commission *v.* Federal Republic of Germany [1991] ECR I-883. The European Commission initiated action against Germany when the Lower Saxony regional authorities, in strengthening sea defences, disturbed birds and their habitat in the Leybucht–Dykes SPA. Germany (supported by the UK) had argued that the duty to protect the site was qualified by the need to take account of economic and recreational considerations.
32 When planning permission is revoked in the UK, compensation is payable: in the case outlined it was to be underwritten by the Secretary of State. This is clearly crucial, since otherwise local authorities can be reluctant to bring such actions.
33 In England, for example, the number of SSSIs subject to actual loss or damage declined sharply from 324 (6 per cent of the total) in 1990 to 65 in 1996/97 (English Nature 1997c). The area covered by SSSIs (in Great Britain) increased by 45 per cent between 1985 and 1997.
34 Irrecoverable damage to SSSIs in 1996/97 included the bisecting of woodland at Snelsmore Common by the Newbury by-pass and the loss of an area of wetland in Chichester Harbour to coastal defence works (English Nature 1997c). More generally, the failure to integrate nature conservation and biodiversity objectives with flood defence functions has been a particular source of criticism of the Environment Agency (ETRAC 2000).
35 In its response to the DETR consultation paper on SSSIs (DETR 1998i), English Nature (1998c: 1) acknowledged that 27 per cent of SSSI management units remained in 'unfavourable management', largely as result of neglect, or failure to implement appropriate management to maintain the features of interest for which the site had been designated.
36 In this case, the RSPB sought a judicial review of the Secretary of State's decision; the matter was ultimately referred to the European Court of Justice (see Miller 1999b for a more detailed account).
37 Confirming the difficulties of being proactive, the survey found that only in 0.7 per cent of cases did planning conditions seek to provide genuine enhancement, and that conditions were usually restrictive rather than positive.
38 Friends of the Earth Cymru, quoted in the *Western Mail*, 15 May 1997, p. 9.
39 In determining the boundaries of the Lower Severn Estuary SPA, the government excluded the Taff/Ely SSSI on the grounds that, in itself, it supported less than 1 per cent of the international population of wading birds necessary to determine such sites. This contradicted the advice of the then Nature Conservancy Council and set aside the argument of conservationists that the SSSI was an integral part of the system of habitats available to bird populations. Economic requirements – to redevelop Cardiff Bay – were also put forward as a reason (see Welsh Office 1991).
40 For example, by an RSPB spokesperson on BBC Radio 4's *Today* programme, 24 May 1999.
41 Application-specific bargaining over impacts and compensation is almost invariably a secretive process, to which the developer has privileged access. Most of what passes for 'environmental compensation' has accommodated the interests of developers more successfully than those of conservationists (Cowell 1997).
42 The developer (ECC Ball Clays) had argued that translocation of the threatened SSSI to a nearby field would provide adequate mitigation, claims that were challenged by English Nature.

7 Distributing development: sustainability and equity in minerals planning

1 As Middleton (1999: 266–7) points out, Agricola, writing in the sixteenth century, felt that the environmental costs of aggregates extraction were worth it: 'if there were no metals, men would face a horrible and wretched existence in the midst of wild beasts; they would return to the acorns and fruits and berries of the forest'.

2 They also present challenges to those seeking to write succinctly about changes to 'national' policy. The Department of the Environment, Transport and the Regions has become the main government department for planning in England only. Until the early 1990s, statements of 'national' minerals policy generally applied to England and Wales, being produced by the Department of the Environment with some input from the Welsh Office. Scotland always had a legally distinct planning system and prepares its own planning policy guidance. Although we refer at times to 'the UK government', we do not include in this chapter any reference to minerals planning in Northern Ireland, which has its own institutional arrangements.

3 In the vernacular sense of a (purely selfish) desire to resolve immediate environmental threats by displacing them to someone else's backyard, while accepting benefits from the type of development in question.

4 Blowers and Leroy (1994) warn against the thought that communities heavily burdened with environmentally degrading activities would automatically be relieved.

5 Responsibility for minerals planning resides with county councils and national park authorities rather than lower-tier district councils, except where no county councils exist, when the function is allocated to London boroughs, metropolitan district councils and unitary authorities.

6 Details of the forecasting methodologies used in the early 1990s are set out in DoE 1994d, Annex C.

7 During the 1990s, the NCG was chaired by the Department of the Environment and included the chairs of the RAWPs, representatives of minerals producers, the DoE minerals and construction divisions, the Welsh Office (now the National Assembly for Wales), the Scottish Office (now the Scottish Executive) and the Ministry for Agriculture Fisheries and Food (DoE 1994d: Appendices A, B and D). Following a review, the NCG has extended its membership to include English Nature, the Countryside Agency, representatives of marine dredging interests and aggregates users. At the time of writing, RAWPs only operate in England and Wales, drawing their membership from minerals planning authorities in each region, the industry trade associations, central government (Department of the Environment, Transport and the Regions in England, or the National Assembly for Wales) and 'other interested parties' (DoE 1994d: para D4).

8 Wales was included within the strategic national minerals planning framework completed in 1989 and was integral to early drafts of the 1994 guidance, MPG 6, before being removed from the final version. Its departure is attributable to the then Secretary of State for Wales (a central government appointment), John Redwood, who sought to promote a more *laissez-faire* approach to planning in the principality.

9 For England, the government indicated that 3.11 billion tonnes of aggregate could come from domestic 'land-won provision' over the period 1992–96, 320 million tonnes from marine-dredged sand and gravel, 530 million tonnes from

secondary and recycled sources, 160 million tonnes from Wales and 160 million tonnes from 'imports from outside England and Wales' (DoE 1994d: 9).

10 These include initiatives to promote more resource-efficient building techniques with the construction industry (DETR 2000g) and to incorporate waste-based aggregates into construction specifications. Planners remain involved in locational aspects, such as finding sites for demolition waste crushers.

11 Inert wastes, such as construction materials, are classified as 'inactive' and therefore attract a lower rate of tax than 'active' wastes. The efficacy of the tax is far from clear, however, with some building waste being diverted from landfill into other low-grade uses like landscaping.

12 Acting as the main UK trade association for the quarrying and construction materials sectors.

13 Although one should note that north-west England, traditionally a region of heavy industry, has also been a major importer of primary aggregates.

14 Inspectors at both the Berkshire and West Sussex inquiries concurred with national policy presumptions in favour of conservation in nationally designated sites such as Areas of Outstanding Natural Beauty, provided that local planning policies did not wholly preclude the scope for companies to make applications for minerals extraction in these areas.

15 Indeed, despite expressed concern to avoid 'exporting unsustainability', the minerals plans of Berkshire and West Sussex both assumed an ability to meet a significant proportion of their own aggregates demand from beyond their borders.

16 Such judgements are also reflected in government research investigating the appropriate level for the aggregates tax (London Economics 1998). In the first phase of the research, local people expressed an average 'willingness to accept' compensation for the impacts of sand and gravel working of £9.50 per tonne. This was in densely populated lowland areas. 'External costs' fell to only 18 pence per tonne for the Glensanda coastal superquarry in a remote part of rural Scotland. This illustrates the difficulties discussed in Chapter 4 of capturing certain values through economic valuation techniques, and the vital importance of political scale in giving presence to certain kinds of values – only people living within five miles of extraction sites were interviewed.

17 And from some rural roads around existing quarries to the environs of exporting and importing wharf facilities that will be needed to handle the large volumes of rock that superquarries produce.

18 Hereafter referred to as Redland but since taken over by French company Lafarge and trading as Lafarge Redland Aggregates Ltd.

19 For further discussion see the Reporter's report (Pain 1999), Scottish Ministers' decision letter (Firth 2000) and other accounts (Link Quarry Group 1996; Mackenzie 1998; Martin and Abercrombie 1995; Owens and Cowell 1996).

20 Redland's economic witnesses drew upon the fact that the Western Isles had been institutionalised as 'peripheral' in social, economic and geographical terms, with a *per capita* GDP only 52 per cent of the UK average (Bailey and Galloway 1993: 152).

21 Western Isles Islands Council until local government reorganisation in 1996.

22 The Inquiry Reporters Unit is the Scottish equivalent of the Planning Inspectorate in England and provides Reporters to chair planning inquiries. In Scotland since devolution, applications that have been called in for a public inquiry come before the Scottish Executive (in practice, before the minister with responsibility for planning), but decisions are issued in terms of 'Scottish Ministers' – the preferred collective phrase of the Executive's ministerial team.

23 Although Redland has appealed against the decision.

24 Except in a few parts of southern England, where marine-dredged aggregates and crushed rock from other countries are landed (noting that most marine-dredged material comes from the UK continental shelf: thus it is 'offshore' but not, technically speaking, an import).

25 Not least by British minerals interests, with half an eye on protecting domestic markets (Pollock 1999; Searle 1975).

26 Falling demand in the UK and Germany ultimately caused the company to abandon the project.

27 And detected at different scales: in debates about the location of exporting quarries in Scotland and Norway; across the former Communist states of Central Europe, and also more locally in England, where some minerals planning authorities have faced dilemmas about whether to concentrate sand and gravel production in certain areas or seek to open up new areas of supply (Murdoch and Marsden 1995).

28 The fate of places economically dependent on producing raw materials also falls outside its compass. As we noted in Chapter 1, theories of ecological modernisation also say relatively little about less material concepts of environmental value (Mol 1996), including the spiritual, aesthetic and intrinsic qualities of the non-human world, which have proved demonstrably important in conflicts over 'eco-efficient' superquarries.

29 Nor are all desirable goals appropriately captured by justice. As we discussed in Chapter 3, the conservation of wildlife or beautiful landscapes might be addressed through ideas of virtue (Dobson 1998; O. O'Neill 1997).

30 Pressure of space prevents our dwelling on the social inequalities, and unequal access to the decision-making process, that prevails *within* territorially defined communities.

31 Harvey (1973: 116), as cited in Hay (1995).

32 Searle (1980), writing about a proposed copper mine in the Snowdonia National Park in the 1970s, observed these arguments being made by some local politicians. Distributive concerns also arise from exploiting resources: at Lingerbay, much discussion centred on the creation of a community trust fund, with Redland contributing a sum per tonne of rock extracted.

33 Concepts of 'environmental space' give more explicit attention to distributive justice in the consumption of benefits from environmental capital, advancing a core principle of equal *per capita* entitlement to the Earth's sustainable productive capacity (Carley and Spapens 1998; Friends of the Earth Europe 1995). Actual ecological footprints can be compared with these entitlements, generally showing that developed nations occupy far more than a fair share of environmental space.

34 This approach has commonalities with capital-based frameworks if interpreted in ways that would oblige us to ensure that all environments meet minimum quality requirements, based on human health and safety. Another approach, discussed in Chapters 3 and 6, would be to allow differential outcomes between regions only where that avoided systematic and gratuitous injury (O. O'Neill 1996) – the matter of gratuitousness calling into account whether potentially damaging forms of development were needed.

35 This presents a major difficulty for those who believe that the distributive impacts of capacity-based planning can be addressed by requiring all local authorities to evaluate their environmental capacity on some standard methodological basis.

36 Claims to 'local autonomy' and 'community' can be deployed instrumentally to de-legitimise non-local perspectives, whether environmental or economic. Witness, for example, Byron (1995: 9) objecting to 'people who will go

anywhere in the country to do their moral hectoring'; 'young environmental advocates from London and other English cities have belatedly travelled to Harris to try to stop the [Lingerbay superquarry] development, and to "educate" the islanders about what is "really" in their interests'.

37 It has long been recognised, for example, that conferring autonomy on lower tiers of government involves a risk that legitimate 'higher-level' needs might not be met (although, as we have already argued, how such 'needs' should be defined is itself a point of contention). Nor is resource autonomy a straightforward idea: a policy agenda that required the need for aggregates to be met locally would create profound problems for urban centres, since, by definition, their 'ecological footprint' extends beyond their physical boundaries (W. Rees 1999).

38 In this respect, making progress towards sustainability faces a number of profound dilemmas evident in rethinking systems of governance more generally. European integration, British political devolution and the balance between national and local government everywhere are just some of the alternative contexts we might name. 'The environment is but one collective affair' (Munton 1999: 20).

39 There is always the risk of collapsing discourses of value into hierarchies of scale. In planning, for example, categorising values as 'personal' or 'local' is a common means of counterposing them unfavourably with the 'universal' and the 'expert' (Beynon *et al.* 2000).

40 The government has embarked on a process of renegotiating old mining rights. The 1995 Environment Act introduced provisions for mandatory fifteen-year reviews of planning conditions for all existing minerals workings (extending earlier powers to review and apply conditions to Interim Development Order permissions dating from before 1947). This seeks to ensure that planning conditions for the operation and rehabilitation of long-term quarries keep up to date with the advance of environmental standards (DETR 2000f: para 2.6). However, where new conditions would reduce the volume of reserves that could be extracted, then compensation is payable to the consent holder.

41 See, for example, Blowers 1993a; London Economics 1999; Low and Gleeson 1998; Shrader-Frechette 1985. Practice has been more *ad hoc*, but the impending introduction of a 'sustainability fund' using aggregates tax revenues provides a potential vehicle. Restoring 'orphaned' extraction sites might be a justifiable use of the fund.

42 Davoudi *et al.* (1997) describe how Lancashire County Council, in the northwest of England, engaged in a number of participatory and evaluation processes around the environmental dimension of its structure plan. Yet minerals interests still dominated consultative exercises about the new minerals plan, and demand management (in the sense of trying to reduce demand) was given little weight.

43 For example, it might allow more space to focus sustainability debates on the relative significance and substitutability of different categories of need, perhaps recognising that the impacts of extracting sand and gravel from intensely farmed land in southern England can more often be acceptably remediated than the environmental, economic and cultural legacy of new, large hard rock quarries.

44 Similar political sensitivities are negotiated in Principle 21 of the Stockholm Declaration:

> States have, in accordance with the charter of the United Nations and the principles of international law, the sovereign right to exploit their own resources … and the responsibility to ensure that activities within their

jurisdiction or control do not cause damage to the environment of other states or areas beyond the limits of national jurisdiction.

45 One should still look beyond the rhetoric. Although the government seems to have displaced primary aggregates from the policy core, primary materials still dominate (about 80 per cent) in terms of actual tonnage.

46 This has also precipitated a split in the representation of the industry, with the British Aggregates Association (BAA) being formed in 1999 from disaffected members of the QPA. Encompassing mostly smaller producers, the BAA was unhappy at the cost of the 'new deal' package that the QPA proposed as an alternative to the aggregates tax, and at the way the QPA was dominated by 'the majors', which were seeking to reduce capacity and acting to squeeze out other operators (*Mineral Planning* 2000). In an intriguing alignment of interests, BAA representatives have criticised the concept of Scottish coastal superquarries as 'selling the family silver', extolling the social and environmental virtues of small-scale local quarries instead (Durward 2000).

8 Conclusions and reflections

1 It is an interesting feature of land-use planning that a pro-development decision is final for the environment when irreplaceable features will be lost, but a refusal of planning permission may provide only temporary respite: there is nothing to stop a development proposal reappearing in a slightly different guise.

2 See, for example, paragraph 36 of Guidance on General Policy and Principles (DoE 1997a), which begins: 'The Government recognises and upholds the rights of property and the privileges of ownership'. Economic development and competitiveness have a high profile throughout the document.

3 To conclude that planning is insufficient is not to suggest that planning policies are inevitably futile, which can become an excuse for inaction: planning can contribute to a reframing of issues over time, as we have shown.

BIBLIOGRAPHY

1000 Friends of Oregon (1997) *Making the Connections: A Summary of the LUTRAQ Project*, Vol. 7: *Integrating Land Use and Transportation Planning for Livable Communities*, Portland, Oregon: 1000 Friends of Oregon.

AA (Automobile Association) (undated) *Transport and the Environment: A Policy Statement*, Basingstoke, Hants: AA Public Policy Department.

Acheson, D. (chair) (1998) *Report of the Independent Inquiry into Inequalities in Health*, London: Stationery Office.

Adams, J. (1981) *Transport Planning: Vision and Practice*, London: Routledge & Kegan Paul.

Adams, W. (1990) *Green Development: Environment and Sustainability in the Third World*, London: Routledge.

—— (1996) *Future Nature*, London: Earthscan.

Advisory Group on Citizenship (1998) *Education for Citizenship and the Teaching of Democracy in Schools*, final report, London: Qualifications and Curriculum Authority.

Agricola, G. (1950) *De Re Metallica* (first edition 1556), translated by H.C. Hoover and L.H. Hoover, New York: Dover Publications.

Agyeman, J. (2000) *Environmental Justice: From the Margins to the Mainstream*, TCPA Tomorrow Series, Paper 7, London: TCPA.

Alden, J. and Boland, P. (eds) (1996) *Regional Development Strategies: A European Perspective*, London: Jessica Kingsley.

Allmendinger, P. and Thomas, H. (eds) (1998) *Urban Planning and the British New Right*, London: Routledge.

Arup Economics and Planning (1995) *Environmental Capacity: A Methodology for Historic Cities*, final report, Chester: Cheshire County Council.

Association of District Councils (1990) *UK Transport: A Balanced Approach*, London: Association of District Councils.

Association of Metropolitan Authorities (1990) *Changing Gear: Urban Transport Policy into the Next Century*, London: Association of Metropolitan Authorities.

Attfield, R. (1987) *A Theory of Value and Obligation*, London: Croom Helm.

Attfield, R. and Dell, K. (1996) *Values, Conflict and the Environment* (second edition), Aldershot: Avebury.

Audit Commission (1999) *Listen Up! Effective Community Consultation*, management paper, Abingdon, Oxfordshire: Audit Commission.

Babtie Group Ltd (for the Royal County of Berkshire) (1993) *Proof of Evidence: General Philosophy of the Plan: The Council's Case*, public inquiry into the Draft Replacement Minerals Local Plan for Berkshire, Document BCC/2a, Reading: Babtie Public Services Division.

Baden, J. and Geddes, P. (2000) *Saving a Place: Endangered Species in the 21st Century*, Aldershot: Ashgate.

Bagehot, W. (1856) 'Dull government', *Saturday Review* 1, 16 (16 February): 287–8.

Bailey, S. and Galloway, S. (1993) 'Financial crisis in the Western Isles', *Local Government Studies* 19: 149–55.

Bain, C., Dodd, A. and Pritchard, D. (1990) *A Study of Development Plans in England and Wales*, RSPB Planscan, Conservation Topic Paper No 28, Sandy, Bedfordshire: RSPB.

Banister, D. (ed.) (1998) *Transport Policy and the Environment*, London: E. & F.N. Spon.

Banister, D. and Button, K. (1992) *Transport, the Environment and Sustainable Development*, London: E. & F.N. Spon.

Banner, M. (1999) *Christian Ethics and Contemporary Moral Problems*, Cambridge: Cambridge University Press.

Barry, B. (1990) *Political Argument* (second edition), New York: Harvester Wheatsheaf.

Bartholomew, K. (1995) 'A tale of two cities', *Transportation* 22: 273–93.

Bartlett, R. and Kurian P. (1999) 'The theory of environmental impact assessment: implicit models of policy making', *Policy and Politics* 27, 4: 415–33.

Barton, H., Davis, G. and Guise, R. (1995) *Sustainable Settlements: A Guide for Planners, Designers and Developers*, Bristol: University of the West of England and Luton: Local Government Management Board.

Bateman, I. (1991) 'Social discounting, monetary evaluation and practical sustainability', *Town and Country Planning* 60, 1: 174–6.

Beatley, T. (1989) 'Environmental ethics and planning theory', *Journal of Planning Literature* 4, 1: 1–32.

—— (1994) 'Environmental ethics and the field of planning', in H. Thomas (ed.) *Values and Planning*, Aldershot: Ashgate, 12–37.

Beattie, C. and Longhurst, J. (1999) 'Progress of air quality management in urban regions of England', paper presented at the National Society for Clean Air Conference, *Environmental Protection*, 25–28 October.

Beckerman, W. (1994) '"Sustainable development": is it a useful concept?' *Environmental Values* 3, 3: 191–209.

—— (1995) *Small is Stupid: Blowing the Whistle on the Greens*, London: Duckworth.

Benton, T. (1993) *Natural Relations: Ecology, Animal Rights and Social Justice*, London: Verso.

Berlin, I. (1969) 'Two concepts of liberty', in *Four Essays on Liberty*, Oxford: Oxford University Press.

Berntsen, T. (1994) *Environmental Policy Statement to the Storting*, 11 April, Oslo: Miljøverndepartmentet.

Beynon, H., Cox, A. and Hudson, R. (2000) *Digging Up Trouble: The Environment Protest and Opencast Mining*, London: Rivers Oram Press.

Bina, O. (2000) (for ERM (Environmental Resources Management)) 'State of the art on TEN-T and SEA', paper presented at the European Forum on Integrated

Environmental Assessment Workshop, *Trans-European Networks*, Institute for European Environmental Policy, London, 7–8 September (available from ERM).

Bina, O. and Vingoe, J. (2000) *Strategic Environmental Assessment in the Transport Sector: An Overview of Legislation and Practice in EU Member States*, report prepared by Environmental Resources Management (ERM) for the European Commission, London: ERM.

Bina, O., Briggs, B. and Bunting, G. (1995) *The Impact of Trans-European Networks on Nature Conservation: A Pilot Project*, Sandy, Bedfordshire: RSPB.

Bishop, K. (1996) 'Planning to save the planet? Planning's green paradigm', in M. Tewdwr-Jones (ed.) *British Planning Policy in Transition*, London: UCL Press, 205–19.

Bishop, K., Phillips, A. and Warren, L. (1995) 'Protected for ever? Factors shaping the future of protected areas policy', *Land Use Policy* 12, 4: 291–305.

Bishop, R. (1978) 'Endangered species and uncertainty: the economics of a safe minimum standard', *American Journal of Agricultural Economics* 60, 1: 10–18.

Blair, T. (1999) 'Foreword', in DETR *A Better Quality of Life*, London: Stationery Office.

Blake, J. (1999) 'Method or madness: a contextual approach to researching environmental values', unpublished PhD thesis, Department of Geography, University of Cambridge.

Blowers, A. (1984) *Something in the Air: Corporate Power and the Environment*, London: Harper & Row.

—— (ed.) (1993a) *Planning for a Sustainable Environment*, a report by the Town and Country Planning Association, London: Earthscan.

—— (1993b) 'The time for change', in A. Blowers (ed.) *Planning for a Sustainable Environment*, London: Earthscan, 1–18.

—— (1997) 'Environmental planning for sustainable development: the international context', in A. Blowers and B. Evans (eds) *Town Planning into the 21st Century*, London: Routledge, 33–53.

—— (1999) 'Nuclear waste and landscapes of risk', *Landscape Research* 24, 3: 241–64.

—— (2000) 'Ecological and political modernisation: the challenge for planning', *Town Planning Review* 71, 4: 371–93.

Blowers, A. and Evans, B. (eds) (1997) *Town Planning into the 21st Century*, London: Routledge.

Blowers, A. and Glasbergen, P. (1995) 'The search for sustainable development', in Blowers, A. and Glasbergen, P. (eds) *Environmental Policy in an International Context: Perspectives on Environmental Problems*, London: Edward Arnold, 163–84.

Blowers, A. and Leroy, P. (1994) 'Power, politics and environmental inequality: a theoretical and empirical analysis of "peripheralisation"', *Environmental Politics* 3: 197–228.

Boehmer-Christiansen, S. and Skea, J. (1991) *Acid Politics*, New York: Belhaven Press.

Booth, P. (1996) *Controlling Development: Certainty and Discretion in Europe, the USA and Hong Kong*, London: UCL Press.

Borgström Hansson, C. and Wackernagel, M. (1999) 'Rediscovering place and accounting space: how to re-embed the human economy', *Ecological Economics* 29: 203–13.

Boulding, K. (1966) 'The economics of the coming Spaceship Earth', in H. Jarrett (ed.) *Environmental Quality in a Growing Economy*, Baltimore: Resources for the Future, 3–14.

Bouwer, K. (1993) 'Integration of regional environmental planning and physical planning in the Netherlands', paper presented at the Institute of British Geographers Conference, January, The Netherlands: Department of Environmental Policy Sciences, University of Nijmegen.

—— (1994) 'The integration of regional environmental planning and physical planning in the Netherlands', *Journal of Environmental Planning and Management* 37, 1: 107–16.

Bowers, J. (1997) *Sustainability and Environmental Economics: An Alternative Text*, Harlow: Longman.

Box, J. and Harrison, C. (1993) 'Natural spaces in urban places', *Town and Country Planning* 62: 231–5.

Bramley, G. and Watkins, C. (1995) *Circular Projections: Household Growth, Housing Development and the Household Projections*, London: CPRE.

Breheny, M. (1995) 'The compact city and transport energy consumption', *Transactions of the Institute of British Geographers* NS 20, 2: 81–101.

Breheny, M. and Hooper, A. (eds) (1985) *Rationality in Planning*, London: Pion.

Breheny, M. and Rookwood, R. (1993) 'Planning the sustainable city region', in A. Blowers (ed.) *Planning for a Sustainable Environment*, London: Earthscan, 150–89.

British Roads Federation (1999) 'Traffic growth reaches ten year high', press release 19/99, 6 August, London: BRF.

Brooke, C.E. (1996) *Natural Conditions – A Review of Planning Conditions and Nature Conservation*, Sandy, Bedfordshire: RSPB.

Brown, P. (2000) 'Environmental battleground: Scottish island quarry rejected', *Guardian* 4 November: 5.

Brown, V. (1998) 'Ground-truthing ecologically sustainable development', in S. Buckingham-Hatfield and S. Percy (eds) *Constructing Local Environmental Agendas: People, Places and Participation*, London: Routledge, 140–55.

Bruff, G. and Wood, A. (1995) 'Sustainable development in English metropolitan district authorities: an investigation using unitary development plans', *Sustainable Development* 3: 9–19.

Brugman, J. (1997) 'Is there method in our measurement? The use of indicators in local sustainable development planning', *Local Environment* 2, 1: 59–72.

Brundell, M. (1994) *Inspector's Report of the Inquiry into the Replacement Minerals Local Plan for Berkshire*, 21 September to 16 November 1993, Bristol: Planning Inspectorate.

Buchanan, C. (1963) *Traffic in Towns*, London: HMSO.

Buckingham-Hatfield, S. (1997) 'Public participation in Local Agenda 21: the usual suspects', in P. Kivell, P. Roberts and G. Walker (eds) *Environment, Planning and Land Use*, Aldershot: Ashgate, 208–19.

Burney, J. (2000) 'Is valuing nature contributing to policy development?' *Environmental Values* 9, 4: 511–20.

Byron, H. (2000) *Biodiversity Impact. Biodoversity and Environmental Impact Assessment: A Good Practice Guide for Road Schemes*, Sandy, Bedfordshire: RSPB, WWF UK, English Nature and the Wildlife Trusts.

Byron, R. (1995) 'Introduction', in R. Byron (ed.) *Economic Futures on the North Atlantic Margin*, Aldershot: Ashgate, 1–15.

CAG (1998) *Environmental Capital* (newsletter), autumn, London: CAG Consultants Ltd.

CAG and Land Use Consultants (1997) *What Matters and Why? Environmental Capital: A New Approach – A Provisional Guide*, report prepared for the Countryside Commission, English Nature, English Heritage and the Environment Agency, Cheltenham: Countryside Commission.

—— (1998) *Environmental Capital and Transport Appraisal*, unpublished report for the Countryside Commission, Cheltenham: Countryside Commission.

—— (1999) *Environmental Capital: Report of the Experience of the Pilot Projects*, Cheltenham: Countryside Agency (Planning Branch).

Caldwell, L. and Schrader-Frechette, K. (1993) *Policy for Land: Law and Ethics*, Lanham, Maryland: Rowman & Littlefield.

Callicott, J. (1989) *In Defence of the Land Ethic; Essays in Environmental Philosophy*, Albany, NY: State University of New York Press.

Callies, D. (1999) 'An American perspective on UK planning', in B. Cullingworth (ed.) *British Planning: 50 Years of Urban and Regional Policy*, London and New Brunswick, NJ: Athlone Press.

Campbell, H. and Marshall, R. (1999) 'Ethical frameworks and planning theory', *International Journal of Urban and Regional Research* 23, 3: 464–78.

—— (2000) 'Public involvement and planning: looking beyond the one to the many', *International Planning Studies* 5, 3: 321–44.

Carley, M. and Spapens, P. (1998) *Sharing the World, Sustainable Living and Global Equity in the 21st Century*, London: Earthscan.

Carter, N. and Darlow, A. (1997) 'Local Agenda 21 and developers: are we better equipped to build a consensus in the 1990s?' *Planning Practice and Research* 12, 1: 45–57.

Cartledge, B. (ed.) (1996) *Transport and the Environment*, the Linacre Lectures 1994/95, Oxford: Oxford University Press.

Cartwright, L. (1997) 'The implementation of sustainable development by local authorities in the South East of England', *Planning Practice and Research* 12, 4: 337–47.

CBI (Confederation of British Industry) (1995) *Transport Brief*, London: CBI.

—— (1998) *Roads to the Market: Economic Instruments in an Integrated Transport Policy*, London: CBI.

CBI and RICS (Royal Institute of Chartered Surveyors) (1992) *Shaping the Nation*, report of the planning task force, London: CBI.

CEC (Commission of the European Communities) (1992a) *Towards Sustainability: A New European Programme of Policy and Action in Relation to the Environment and Sustainable Development*, COM(92)23 final, Brussels: CEC.

—— (1992b) *The Future Development of the Common Transport Policy: A Global Approach to the Construction of a Community Framework for Sustainable Mobility*, COM(92)494 final, Brussels: CEC.

—— (1996) *Evaluation of the Performance of the EA Process*, final report, Volumes 1 and 2, Luxembourg: Office for Official Publications of the European Commission.

—— (1997) 'Proposal for a Council Directive on the Assessment of the Effects of Certain Plans and Programmes on the Environment', *Official Journal of the European Communities* 97/C 129/08, 25 April: 14–18.

—— (1999a) *Towards Balanced and Sustainable Development of the Territory of the European Union*, Luxembourg: Office for Official Publications of the European Commission.

—— (1999b) *Amended Proposal for a Council Directive on the Assessment of the Effects of Certain Plans and Programmes on the Environment*, COM(99)73, Luxembourg: Office for Official Publications of the European Commission.

—— (1999c) *Global Assessment of the Fifth Environmental Action Programme*, COM(99)543 final, Luxembourg: Office for Official Publications of the European Commission.

CfIT (Commission for Integrated Transport) (1999) *National Road Traffic Targets*: advice by CfiT, London: DETR (and at http://www.detr.gov.uk/cfit/index.htm).

Chambers, R. (1999) 'Will UK National Park reforms act as beacons for raised environmental performance – or are they a flash in the pan?' in P. Fuchs, M. Smith and M. Arthur (eds) *Mineral Planning in Europe*, Nottingham: Institute of Quarrying, 114–25.

Christoff, P. (1996) 'Ecological modernisation, ecological modernities', *Environmental Politics* 5, 3: 476–500.

City and County of Cardiff (1997) *Issues for the Unitary Development Plan*, Cardiff: Planning Department, City and Council of Cardiff.

City of York (1993) *Local Plan*, York: York City Council.

Clark, J., Burgess, J. and Harrison, C. (2000) '"I struggled with this money business" – respondents' perspectives on contingent valuation', *Ecological Economics* 33, 1: 45–62.

Clark, M., Burrell, P. and Roberts, P. (1993) 'A sustainable economy', in A. Blowers (ed.) *Planning for a Sustainable Environment*, London: Earthscan, 131–49.

Clark, S. (1977) *The Moral Status of Animals*, Oxford: Oxford University Press.

Clifford, S. (1997) 'Local distinctiveness', *Nature's Place* 15: 4–5.

Coase, R. (1960) 'The problem of social cost' *Journal of Law and Economics* 3: 1–44.

Cole, H. (2000) 'Regional plans just don't add up', *Planning* 8 September, 11.

Cole, L. (1997) 'A capital idea – a new way of looking at the environment', *Nature's Place* 16: 3–5.

Collis, I., Heap, J. and Jacobs, M. (1992) 'Strategic planning and sustainable development', paper prepared for English Nature, Peterborough: English Nature.

Connell, B. (1999) 'Accommodating development: a view from West Sussex – environmental capacity and strategic development choices', paper presented to the Planning, Space and Sustainability Seminar II: Concepts and Tools, Cardiff University, 13 October, Chichester: West Sussex County Council.

Cordrey, L. (ed.) (1997) *Action for Biodiversity in the South West: A Series of Habitat and Species Plans to Guide Delivery*, Exeter: RSPB South West England Office.

Counsell, D. (1998) 'Sustainable development and structure plans in England and Wales: a review of current practice', *Journal of Environmental Planning and Management* 41, 2: 177–94.

—— (1999a) 'Making sustainable development operational', *Town and Country Planning* 68, 4: 131–3.

—— (1999b) 'Sustainable development and structure plans in England and Wales: operationalizing the themes and principles', *Journal of Environmental Planning and Management* 42, 1: 45–61.

—— (1999c) 'Attitudes to sustainable development in the housing capacity debate: a case study of the West Sussex Structure Plan', *Town Planning Review* 70, 2: 213–29.

Countryside Agency (1999) *Renewable Energy Development: The Role of the Countryside Agency*, Countryside Agency Board Paper AP99/50 (by R. Minter), Cheltenham: Countryside Agency.

Countryside Commission (1988) *Annual Report*, Cheltenham: Countryside Commission.

—— (1990) *Annual Report*, Cheltenham: Countryside Commission.

—— (1993) *Sustainability and the English Countryside*, Cheltenham: Countryside Commission.

—— (1996) *England's Countryside: The Role of the Planning System*, CCP 496, Cheltenham: Countryside Commission.

—— (1997) 'Government urged to consider by-pass alternatives for Salisbury', news release NR/97/8, 17 February, Cheltenham: Countryside Commission.

—— (1998) *Planning for Countryside Quality*, CCP 529, Cheltenham: Countryside Commission.

Countryside Commission, English Heritage and English Nature (1993) *Conservation Issues in Strategic Plans*, CCP 420, Cheltenham: Countryside Commission.

—— (1996) *Conservation Issues in Local Plans*, CCP 485, London: English Heritage.

County Planning Officers' Society (1993) *Planning for Sustainability*, CPOS, Winchester: Hampshire County Council.

Cowell, R. (1997) 'Stretching the limits: environmental compensation, habitat creation and sustainable development', *Transactions of the Institute of British Geographers* NS 22: 392–406.

—— (2000a) 'Environmental compensation and the mediation of environmental change: making capital out of Cardiff Bay', *Journal of Environmental Planning and Management* 43, 5: 689–710.

—— (2000b) 'Localities and the international trade in aggregates: coastal superquarries on the horizon?' *Geography* 85, 2: 134–44.

Cowell, R. and Murdoch, J. (1999) 'Land use and limits to (regional) governance: some lessons from planning for housing and minerals in England', *International Journal of Urban and Regional Research* 23, 4: 654–69.

Cowell, R. and Owens, S. (1998) 'Suitable locations: equity and sustainability in the minerals planning process', *Regional Studies* 32, 9: 797–811.

Cowell, R., Jehlicka, P., Marlow, P. and Owens, S. (1998) *Aggregates, Trade and the Environment*, London: IUCN UK Committee.

CPRE (Council for the Protection of Rural England) (1991) *Planning Policy Guidance: General Policy and Principles of the Planning System*, London: CPRE.

—— (1995) *The Great Transport Debate: A Submission by CPRE*, London: CPRE.

—— (1997) *Household Growth: Where Shall We Live? A Response to the Government's Green Paper*, London: CPRE.

—— (1999a) *Sprawl Patrol: How Bad Planning Decisions are Causing Traffic Growth*, London: CPRE.

—— (1999b) *New and Renewable Energy: Prospects for the 21st Century*, response to DTI consultation paper, London: CPRE.

—— (2000a) evidence to Royal Commission on Environmental Pollution Study of Environmental Planning: http://www.rcep.org.uk

—— (2000b) *New Agenda for Sustainable Minerals Planning*, press release, 16 February, London: CPRE.

Crane, D. (1972) *Invisible Colleges*, Chicago: University of Chicago Press.

Crenson, M. (1971) *The Un-Politics of Air Pollution*, Baltimore and London: Johns Hopkins University Press.

Crofts, R. (1998) 'Rural development and the natural heritage', paper presented to the Scottish National Rural Partnership Working Group, 13 February, Edinburgh: Scottish Natural Heritage.

Cullinane, S. and Stokes, G. (1998) *Rural Transport Policy*, Oxford: Elsevier.

Cullingworth, B. (1997) 'British land-use planning: a failure to cope with change?' *Urban Studies* 34, 5–6: 945–60.

Cullingworth, B. and Nadin, V. (1994) *Town and Country Planning in Britain*, London: Routledge.

Daly, H. (1995) 'On Wilfred Beckerman's critique of sustainable development', *Environmental Values* 4, 1: 49–55.

Daly, H. and Cobb, J. (1989) *For the Common Good: Redirecting the Economy towards Community, the Environment and a Sustainable Future*, London: Green Print.

David Tyldesley and Associates (2000) *RSPB Positive Planning Project: Regulation 37 and LBAP Policies in Development Plans*, Edinburgh and Nottingham: David Tyldesley and Associates.

David Tyldesley and Associates (with CAG Consultants) (1994) *Planning for Environmental Sustainability*, Peterborough: English Nature.

Davidoff, P. (1965) 'Advocacy and pluralism in planning', *Journal of the American Institute of Planners* 31, 4: 331–8.

Davies, A. (1998) 'Environmental values and the UK land use planning system', unpublished PhD thesis, Department of Geography, University of Cambridge.

Davies, H., Edwards, D., Hooper, A. and Punter, J. (1989) *Planning Control in Western Europe*, London: HMSO.

Davis, N. (1993) 'Contemporary deontology', in P. Singer (ed.) *A Companion to Ethics*, Oxford: Basil Blackwell, 205–18.

Davoudi, S., Healey, P. and Hull, A. (1997) 'Rhetoric and reality in British structure planning in Lancashire, 1993–1995', in P. Healey, A. Khakee, A. Motte and B. Needham (eds) *Making Strategic Plans: Innovation in Europe*, London: UCL Press, 153–71.

de Jong, B. and de Mulder, E. (1998) 'Construction materials in the Netherlands: resources and policy', in P. Bobrowsky (ed.) *Aggregate Resources: A Global Perspective*, Rotterdam: Balkema, 203–14.

de Schiller, S. (1999) 'Sustainable urban form, environment and climate responsive design', in M. Carmona (ed.) *Globalisation, Urban Form and City Governance*, Delft: Publicatiebureau Bouwkunde.

Dean, L. (1995) *The Use of Indicators in Structure Planning: Tools for the New Environmental Agenda?* Papers in Planning Research No. 158, Cardiff: University of Wales, Department of City and Regional Planning.

Degeling, P. (1995) 'The significance of "sectors" in calls for urban public health intersectoralism: an Australian perspective', *Policy and Politics* 23, 4: 289–301.

Demerrit, D. (1994) 'Ecology, objectivity and critique in writings on nature and human societies', *Journal of Historical Geography* 20, 1: 22–37.

DETR (Department of the Environment, Transport and the Regions) (1997a) *The Application of Environmental Capacity to Land Use Planning*, London: DETR.

—— (1997b) 'Gavin Strang launches new approach on strategic roads with decision reached on twelve accelerated review cases', News Release 176/Transport, 28 July, London: DETR.

—— (1998a) *Modernising Planning*, London: DETR.

—— (1998b) *Planning for the Communities of the Future*, London: HMSO.

—— (1998c) *Planning for Sustainable Development: Towards Better Practice*, London: DETR.

—— (1998d) *Opportunities for Change*, consultation paper on a revised UK strategy for sustainable development, London: DETR.

—— (1998e) *Strategic Environmental Appraisal*, report of an international seminar held in Lincoln, 27–29 May, London: DETR.

—— (1998f) *A New Deal for Transport: Better for Everyone*, Cmnd 3950, London: Stationery Office.

—— (1998g) *A New Deal for Trunk Roads in England*, London: Stationery Office.

—— (1998h) *Places, Streets and Movement*, a companion guide to Design Bulletin 32: *Residential Roads and Footpaths*, London: DETR.

—— (1998i) *Sites of Special Scientific Interest: Better Protection and Management*, consultation paper, London: DETR.

—— (1998j) *Review of the Overall Approach to Planning for the Supply of Aggregates*, London: HMSO.

—— (1998k) *Sustainable Construction*, London: DETR.

—— (1999a) Planning Policy Guidance Note 11 (PPG 11): *Regional Planning*, public consultation draft, London: DETR.

—— (1999b) Revision of Planning Policy Guidance Note 12 (PPG 12): *Development Plans*, public consultation draft, London: DETR.

—— (1999c) *A Better Quality of Life: A Strategy for Sustainable Development for the United Kingdom*, Cmnd 4345, London: Stationery Office.

—— (1999d) *Modernising Planning: A Progress Report*, London: DETR.

—— (1999e) *Modernising Planning: Streamlining the Processing of Major Projects through the Planning System*, London: DETR.

—— (1999f) Revision of Planning Policy Guidance Note 13 (PPG 13*): Transport*, public consultation draft, London: DETR.

—— (1999g) *Quality of Life Counts. Indicators for a Strategy for Sustainable Development for the United Kingdom: A Baseline Assessment*, London: Stationery Office.

—— (1999h) *Sustainable Distribution: A Strategy*, London: Stationery Office.

—— (2000a) *The Air Quality Strategy for England, Scotland, Wales and Northern Ireland*, Cmnd 4548, London: Stationery Office.

—— (2000b) Planning Policy Guidance Note 3 (PPG 3): *Housing*, London: DETR.

—— (2000c) *Good Practice Guide on Sustainability Appraisal of Regional Planning Guidance*, London: DETR.

—— (2000d) *Transport 2010: The Ten-Year Plan*, London: DETR.

—— (2000e) *Tackling Congestion and Pollution*, London: Stationery Office.

—— (2000f) *Planning for the Supply of Aggregates in England: A Draft Consultation Paper*, London: DETR.

—— (2000g) *Building a Better Quality of Life: A Strategy for Sustainable Construction*, London: DETR.

DETR and EFTEC (Economics for the Environment Consultancy) (1998) *Review of the Technical Guidance on Environmental Appraisal*, London: DETR.

Dobson, A. (1998) *Justice and the Environment: Conceptions of Environmental Sustainability and Dimensions of Social Justice*, Oxford: Oxford University Press.

Dodgson, J., Spackman, M., Pearman, A. and Phillips, L. (2000) *Multi-Criteria Analysis: A Manual*, London: DETR.

DoE (Department of the Environment) (1976) *Aggregates: The Way Ahead*, report of the Advisory Committee on Aggregates, London: HMSO.

—— (1980) *Development Control: Policy and Practice*, Circular 22/80, London: HMSO.

—— (1989) Minerals Planning Guidance Note 6 (MPG 6): *Guidelines for Aggregates Provision in England and Wales*, London: DoE.

—— (1991) *Policy Appraisal and the Environment*, London: HMSO.

—— (1992a) Planning Policy Guidance Note 12 (PPG 12): *Development Plans and Regional Planning Guidance*, London: DoE.

—— (1992b) Planning Policy Guidance Note 3 (PPG 3): *Housing*, London: DoE.

—— (1993a) *The Environmental Appraisal of Development Plans: A Good Practice Guide*, London: HMSO.

—— (1993b) Planning Policy Guidance Note 6 (PPG 6): *Town Centres and Retail Development*, London: DoE.

—— (1994a) Planning Policy Guidance Note 9 (PPG 9): *Nature Conservation*, London: DoE.

—— (1994b) *Quality in Town and Country: A Discussion Document*, London: DoE.

—— (1994c) Regional Planning Guidance Note 9 (RPG 9): *Regional Planning Guidance for the South East*, London: DoE.

—— (1994d) Minerals Planning Guidance Note 6 (MPG 6): *Guidelines for Aggregates Provision in England*, London: DoE.

—— (1994e) *Environmental Appraisal in Government Departments*, London: HMSO.

—— (1994f) *Community Involvement in Planning and Development Processes*, London: HMSO.

—— (1995) *Projections of Households in England to 2016*, London: HMSO.

—— (1996a) *Minerals Planning and Supply Practices in Europe*, London: HMSO.

—— (1996b) Minerals Planning Guidance Note 1 (MPG 1): *General Considerations and the Development Plan System*, London: DoE.

—— (1996c) Planning Policy Guidance Note 6 (PPG 6): *Town Centres and Retail Development*, London: DETR.

—— (1997a) Planning Policy Guidance Note 1 (PPG 1, revised): *General Policy and Principles*, London: DoE.

—— (1997b) *The UK National Air Quality Strategy*, Cmnd 3587, produced jointly with the Scottish Office and the Welsh Office, London: Stationery Office.

DoE and DoT (Department of Transport) (1992) *Residential Roads and Footpaths – Layout Considerations*, Design Bulletin 32 (first published 1977), London: DoE/DoT.

—— (1994) Planning Policy Guidance Note 13 (PPG 13): *Transport* London: HMSO.

Don, M. (1999) 'On the verge of perfection', *Observer Magazine*, 8 August: 50.

Donnison, R.D. (1990) *Appeal by British Coal (Opencast Executive): The Shilo North Site*, Inspector's report, Bristol: Planning Inspectorate.

DoT (Department of Transport) (1977) *Report of the Advisory Committee on Trunk Road Assessment*, London: HMSO.

—— (1983) *Manual of Environmental Appraisal* (for trunk roads), London: HMSO.

—— (1989a) *National Road Traffic Forecasts (Great Britain)*, London: HMSO.

—— (1989b) *Roads for Prosperity*, Cmnd 693, London: HMSO.

—— (1996) *Transport, the Way Forward: The Government's Response to the Transport Debate*, Cmnd 3234, London: HMSO.

Downs, A. (1972) 'Up and down with ecology – the "issue-attention cycle"', *Public Interest*, Summer: 38–50.

Downs, P., Gregory K. and Brookes A. (1991) 'How integrated is river basin management?' *Environmental Management* 15, 3: 290–309.

Dreissen, P. and Glasbergen, P. (1995) 'Strategies for network management in an agricultural region: the case of the Gelre valley', in P. Glasbergen (ed.) *Managing Environmental Disputes: Network Management as an Alternative*, Dordrecht: Kluwer, 37–51.

Drysek, J. (1997) *The Politics of the Earth: Environmental Discourses*, Oxford: Oxford University Press.

DTI (Department of Trade and Industry) (2000) *Energy Projections for the UK*, working paper, London: DTI EPTAC Directorate.

Dudley, G. and Richardson, J. (1996a) 'Why does policy change over time? Adversarial policy communities, alternative policy arenas, and British trunk roads policy 1945–95', *Journal of European Public Policy*: 63–83.

—— (1996b) 'Promiscuous and celibate ministerial styles: policy change, policy networks and British roads policy', *Parliamentary Affairs* 49: 566–83.

Dunion, K. (1992) 'Scottish superquarries', *ECOS* 13, 4: 53–4.

Durward, R. (2000) 'Harris superquarry case does not stand up', letter to *The Herald*, 19 September.

Eckersley, R. (1992) *Environmentalism and Political Theory: Towards an Ecopolitical Approach*, London: UCL Press.

ECMT (European Conference of Ministers of Transport) and OECD (Organisation for Economic Co-operation and Development) (1995) *Urban Travel and Sustainable Development*, Paris: OECD Publications Service.

ECOTEC (1993) *Reducing Transport Emissions Through Planning*, London: HMSO.

Elander, I., Lidskog, R. and Johansson, M. (1997) 'Environmental policies and urban planning in Sweden', *European Spatial Research and Policy* 4, 2: 5–35.

Ellerbrock, H. (1999) 'Gravel mining in North Rhine–Westphalia; situation, problems and settlement by regional planning', in P. Fuchs, M. Smith and M. Arthur (eds) *Mineral Planning in Europe*, Nottingham: Institute of Quarrying, 199–202.

Elliot, R. (1993) 'Environmental ethics', in P. Singer (ed.) *A Companion to Ethics*, Oxford: Basil Blackwell, 284–93.

—— (1997) *Faking Nature: The Ethics of Environmental Restoration*, London: Routledge.

Elliott, J. (1999) *Introduction to Sustainable Development* (second edition), London: Routledge.

Ellis, H. (1999) 'Public participation in mineral planning: the collapse of consensus', in P. Fuchs, M. Smith and M. Arthur (eds) *Mineral Planning in Europe*, Nottingham: Institute of Quarrying, 99–113.

Emel, J. and Bridge, G. (1995) 'The Earth as input: resources', in R. Johnston, P. Taylor and M. Watts (eds) *Geographies of Global Change: Remapping the World in the Late Twentieth Century*, Oxford: Basil Blackwell, 318–32.

Emmelin, L. (1998) 'Evaluating environmental impact assessment systems – part 1: theoretical and methodological considerations', *Scandinavian Housing and Planning Research* 15: 129–48.

ENDS (1998a) 'Water abstraction decision deals savage blow to cost–benefit analysis', *ENDS Report* 278, March: 16–18.

—— (1998b) 'A taxing challenge for the quarrying industry', *ENDS Report* 280, May: 2.

—— (1998c) 'DETR considers measures to boost recycled aggregates', *ENDS Report* 280, May: 14–15.

—— (1999a) 'Government's transport wobbles threaten CO_2 target', *ENDS Report* 298, November: 5–6.

—— (1999b) 'Passions stirred over fuel duty's impact on road hauliers', *ENDS Report* 291, April: 37–8.

—— (1999c) 'Air quality management and the art of the possible', *ENDS Report* 288, January: 17–21.

—— (2000a) 'Aggregates tax, 44-tonne lorries in Budget controversies', *ENDS Report* 302, March: 19–21.

—— (2000b) '£180 billion to fix transport or save New Labour?' *ENDS Report* 306, July: 16–19.

—— (2000c) 'Brown spins the government out of a fuel crisis', *ENDS Report* 310, November: 4–5.

English Nature (1993a) *Natural Areas: Setting Nature Conservation Objectives*, consultation paper, Peterborough: English Nature.

—— (1993b) *Position Statement on Sustainable Development*, Peterborough: English Nature.

—— (1995) 'A new breakthrough in planning for nature', *Nature's Place* 9: 15.

—— (1996) *A Space for Nature: Nature is Good for You*, leaflet, Peterborough: English Nature.

—— (1997a) *Position Statement on Roads and Nature Conservation*, Peterborough: English Nature.

—— (1997b) 'Beyond 4000…', *Nature's Place* 16: 11.

—— (1997c) *Sixth Report: 1st April 1996 to 31st March 1997*, Peterborough: English Nature.

—— (1998a) 'Missing Lincs discovered in parish survey', *Nature's Place* 17: 3.

—— (1998b) 'Opportunities for comment', *Nature's Place* 18: 5–6.

—— (1998c) Response to DETR consultation paper: 'Sites of Special Scientific Interest better protection and management', Peterborough: English Nature.

—— (1998d) 'Local nature reserves in the news', *Nature's Place* 17: 12–13.

—— (1999a) 'A new approach to appraising transport projects', *Nature's Place* (Transport Special), March: 4–5.

—— (1999b) 'Guarded welcome by RSPB for the Roads Review', *Nature's Place* (Transport Special), March: 11.

—— (2000) Evidence to Royal Commission on Environmental Pollution Environmental Planning Study: http://www.rcep.org.uk

English Nature and the Countryside Commission (undated) *The Character of England: Landscape, Wildlife and Natural Features*, information leaflet, Peterborough: English Nature and Cheltenham: Countryside Commission.

English Nature, RSPB and the Institute of Terrestrial Ecology (1998) *The Wet Grassland Guide*, Sandy, Bedfordshire: RSPB.

Environment Agency, House of Commons, Environment, Transport and Regional Affairs Select Committee (2000) *Agriculture and the Environment: An Impact Statement Prepared by the Environment Agency*, consultation draft, Bristol: Environment Agency.

ESRC (Economic and Social Research Council) (1999) *The Politics of GM Food: Risk, Science and Public Trust*, Special Briefing 5, Brighton, Sussex: ESRC Global Environmental Change Programme, University of Sussex.

ETRAC (Environment, Transport and Regional Affairs Select Committee) (1999) *The Integrated Transport White Paper*, Vol. 1: *Report and Proceedings*, HC 32-I, Session 1998–99, London: Stationery Office.

—— (2000) *The Environment Agency*, Vol. 1: *Report and Proceedings*, HC 34-I, Session 1999–2000, London: Stationery Office.

European Parliament (2000) *Resolution on the Common Position for Adopting a European Parliament and Council Directive on the Assessment of the Effects of Certain Plans and Programmes on the Environment*, A5–0196–2000, Luxembourg, Office for Official Publications of the European Community.

Evans, B. (1997) 'From town planning to environmental planning', in A. Blowers and B. Evans (eds) *Town Planning into the 21st Century*, London: Routledge, 1–14.

Faber, D. (ed.) (1998) *The Struggle for Ecological Democracy: Environmental Justice Movements in the United States*, London: Guilford Press.

Farmer, A., Skinner, I., Wilkinson, D. and Bishop, K. (1999) *Environmental Planning in the United Kingdom*, report prepared for the Royal Commission on Environmental Pollution, London: Institute for European Environmental Policy.

Farrington, J., Gray, D., Roberts, D. and Martin, S. (1998) 'Rural sustainability and the fuel price escalator', *Town and Country Planning* 67, December: 370–1.

Faulkner, J. (1999) 'Bringing biodiversity to people', *Urban Wildlife News* 16, 4: 12.

Firth, I. (2000) *Town and Country Planning (Scotland) Act 1997. Proposed Extraction, Processing and Transport by Sea of Anorthosite from Land Near Rodel, Isle of Harris, and to Construct the Relevant Facilities*, decision letter, Edinburgh: Scottish Executive Development Department.

Fischer, F. and Forester, J. (eds) (1993) *The Argumentative Turn in Policy Analysis and Planning*, London: UCL Press.

Fitzsimmons, M. (1989) 'The matter of nature', *Antipode* 21: 106–20.

Fletcher, T. and McMichael, A. (1997) *Health at the Crossroads: Transport Policy and Urban Health*, Chichester: John Wiley & Sons.

Floyer-Acland, A. (1990) *A Sudden Outbreak of Common Sense: Managing Conflict Through Mediation*, London: Hutchinson Business Books.

Flyvbjerg, B. (1992) 'Aristotle, Foucault and progressive phronesis: outline of an applied ethics for sustainable development', *Planning Theory* 7–8: 65–83.

—— (1998) *Rationality and Power: Democracy in Practice*, Chicago and London: University of Chicago Press.

Foresight Energy and Natural Environment Panel (2000) *Making Sustainability Count*, consultation paper produced by the Environmental Appraisal Task Force, London: DTI.

Forester, J. (1994a) 'Political judgement and learning about value in transportation planning: bridging Habermas and Aristotle', in H. Thomas (ed.) *Values and Planning*, Aldershot: Ashgate, 178–204.

—— (1994b) 'Bridging interests and community: advocacy planning and the challenges of deliberative democracy', *American Planning Association Journal* 60, 2: 153–8.

—— (1999) *The Deliberative Practitioner: Encouraging Participatory Planning Processes*, Cambridge, Mass.: MIT Press.

Foster J. (ed.) (1997) *Valuing Nature: Economics, Ethics and Environment*, London: Routledge.

Foucault, M. (1982) 'The subject and power', in H. Dreyfus and P. Rabinow (eds) *Michel Foucault: Beyond Structuralism and Hermeneutics*, Brighton: Harvester Press.

—— (1990) *The History of Sexuality*, Vol. 1: *An Introduction*, London: Penguin.

Freudenberg, N. and Steinsapir, C. (1991) 'Not in our backyards: the grassroots environmental movement', *Society and Natural Resources* 4, 3: 235–45.

Friday, L. and Coulston, A. (1999) 'Wicken Fen – the restoration of a wetland nature reserve', *British Wildlife* 11, 1 (October): 37–46.

Friends of the Earth Europe (1995) *Towards a Sustainable Europe*, Brussels: Friends of the Earth Europe.

Gerrard, M. (1994) *Whose Backyard, Whose Risk: Fear and Fairness in Toxic and Nuclear Waste Siting*, Cambridge, Mass.: MIT.

Ghazi, P. (1994) 'Brake now, Minister, or there will be tears', *Observer*, 20 February.

Gibbs, D., Longhurst, J. and Braithwaite, C. (1998) 'Struggling with sustainability: weak and strong interpretations of sustainable development within local authority practices', *Environment and Planning A* 30, 8: 1351–66.

Gilbert, R., Stevenson, D., Girardet, H. and Stren, R. (1996) *Making Cities Work: The Role of Local Authorities in the Urban Environment*, London: Earthscan.

Gillespie, J. and Shepherd, P. (1995) *Establishing Criteria for Identifying Critical Natural Capital in the Terrestrial Environment: A Discussion Paper*, Research Report No. 141, Peterborough: English Nature.

Glasbergen, P. and Driessen, P. (1994) 'New strategies for environmental policy: regional network management in the Netherlands', in M. Wintle and R. Reeve (eds) *Rhetoric and Reality in Environmental Policy: The Case of the Netherlands in Comparison with Britain*, Aldershot: Avebury, 25–40.

Glasgow City Council (1998) *A New Beginning: Glasgow Local Plan Review*, Glasgow: Department of Physical and Economic Regeneration, Glasgow City Council.

Glasson, J. (1999) 'Environmental impact assessment – impact on decisions', in J. Petts (ed.) *Handbook of Environmental Impact Assessment*, Vol. 1, Oxford: Blackwell Science, 121–44.

Glasson, J., Therivel, R. and Chadwick, A. (1994) *Introduction to Environmental Impact Assessment*, London: UCL Press.

Gleeson, B. (2000) 'Devolution and state planning systems in Australia', paper presented at the International Conference: *Devolution and Spatial Planning: Future Directions for Wales and UK Regions?* 27 October, Cardiff: Cardiff University.

Goodin, R. (1989) 'Theories of compensation', *Oxford Journal of Legal Studies* 9, 1: 56–75.

—— (1992) *Green Political Theory*, Cambridge: Polity Press.

—— (1995) *Utilitarianism as a Public Philosophy*, Cambridge: Cambridge University Press.

Goodland, R. and Mercier, J. (1999) *The Evolution of Environmental Assessment in the World Bank: From 'Approval' to 'Results'*, Environment Department Papers No. 67, Washington: World Bank.

Goodpaster, K. (1978) 'On being morally considerable', *Journal of Philosophy* 75: 308–25.

Goodstein, E. (1995) *Economics and the Environment*, Englewood Cliffs, NJ: Prentice Hall.

Goodwin, P. (1993) 'Efficiency and the environment: possibilities of a green–gold coalition', in D. Banister and K. Button (eds) *Transport, the Environment and Sustainable Development*, London: E. & F.N. Spon, 257–69.

—— (1996) 'Transport policy', paper presented at Cambridge Econometrics Annual Conference: *Beyond the Manifestoes – Policies and the Economy after the Election*, Robinson College, Cambridge, 8–9 July, Cambridge: Cambridge Econometrics.

—— (1999) 'Action or inertia? One year on from "A New Deal for Transport"', paper presented at a meeting of the Transport Planning Society, Institution of Civil Engineers, 22 July.

Goodwin, P., Hallett, S., Kenny, F. and Stokes, G. (1991) *Transport: The New Realism*, report to the Rees Jeffreys Road Fund, Oxford: University of Oxford Transport Studies Unit.

Gordon, P. and Richardson, H. (1989) 'Gasoline consumption and cities – a reply', *Journal of the American Planning Association* 55: 342–5.

Government Office for the East of England (2000) *Draft Regional Planning Guidance for East Anglia*, Vol. 2 (modified text incorporating proposed changes by the Secretary of State for the Environment, Transport and the Regions), Government Office for the East of England.

Grant, M. (ed.) (1998) *Encyclopaedia of Planning Law* (first edition 1959, ed. D. Heap) Vol. 1, London: Sweet & Maxwell, 10059–60.

Gregory, R. (1971) *The Price of Amenity: Five Studies in Conservation and Government*, London: Macmillan.

—— (1974) 'The Minister's line: or, the M4 comes to Berkshire', in R. Kimber and J. Richardson (eds) *Campaigning for the Environment*, London: Routledge & Kegan Paul, 103–35.

Grigson, S. (1995) *The Limits of Environmental Capacity*, paper commissioned by the Barton Willmore Partnership and the Housebuilders Federation, London: Housebuilders Federation.

Groom, B. (1999a) 'Planners back Vodafone £120m world HQ', *Financial Times*, 29 April: 14.

—— (1999b) 'Battle lines drawn over strategy for coping with the crowded south-east', *Financial Times*, 18 May: 12.

Grove-White, R. (1991) 'Land use law and the environment', in R. Churchill, J. Gibson and L. Warren (eds) *Law, Policy and the Environment* (special issue of *Journal of Law and Society*), Oxford: Basil Blackwell, 32–47.

—— (1997) 'The environmental valuation controversy: observations on its recent history and significance', in J. Foster (ed.) *Valuing Nature: Economics, Ethics and Environment*, London: Routledge, 21–31.

Habermas, J. (1986) *Autonomy and Solidarity*, London: Verso.

—— (1987) *The Philosophical Discourse of Modernity*, translated from the German by F. Lawrence, Cambridge: Polity Press.

Hajer, M. (1995) *The Politics of Environmental Discourse*, Oxford: Oxford University Press.

Hajer, M. and Kesselring, S. (1999) 'Democracy in the risk society? Learning from the new politics of mobility in Munich', *Environmental Politics* 8, 3: 1–23.

Hall, D., Hebbett, M. and Lusser, H. (1993) 'The planning background', in A. Blowers (ed.) *Planning for a Sustainable Environment*, London: Earthscan, 19–29.

Hall, P.A. (1993) 'Policy paradigms, social learning and the state: the case of economic policy-making in Britain', *Comparative Politics*, April: 275–96.

Hall, P.G. (1980) *Great Planning Disasters*, Harmondsworth: Penguin.

—— (1992) *Urban and Regional Planning* (third edition), London: Routledge.

Hamer, M. (1987) *Wheels within Wheels: A Study of the Road Lobby*, London: Routledge & Kegan Paul.

Hanf, K. and Jansen, A. (eds) (1998) *Governance and Environment in Western Europe*, Harlow: Longman.

Hanley, N. and Spash, C. (1993) *Cost Benefit Analysis and the Environment*, Cheltenham: Edward Elgar.

Hanley, N., Hallett, S. and Moffatt, I. (1990) 'Why is more notice not taken of economists' prescriptions for the control of pollution?' *Environment and Planning A* 22: 1421–39.

Haq, G. (1997) *Towards Sustainable Transport Planning: A Comparison between Britain and the Netherlands*, Aldershot: Avebury.

Hargrove, E. (1992) 'Weak anthropocentric intrinsic value', in M. Oelschlaeger (ed.) *After Earth Day: Continuing the Conservation Effort*, Denton, Texas: University of North Texas Press, 141–69.

Harrison, C. and Burgess, J. (1994) 'Social constructions of nature: a case study of conflicts over the development of Rainham Marshes', *Transactions of the Institute of British Geographers* 19, 3: 291–310.

Harrison, R. (1997) 'Aggregates: the issues on the horizon', *Mineral Planning* 73: 8–11.

Hartwick, J. (1978) 'Investing returns from depleting renewable resource stocks and inter-generational equity', *Economic Letters* 1: 85–8.

Harvey, D. (1973) *Social Justice and the City*, London: Arnold.

—— (1982) *The Limits to Capital*, Oxford: Basil Blackwell.

—— (1996) *Justice, Nature and the Geography of Difference*, Cambridge, Mass., and Oxford: Basil Blackwell.

Hausman, D. and McPherson, M. (1993) 'Taking ethics seriously: economics and contemporary moral philosophy', *Journal of Economic Literature* XXXI, June: 671–731.

Hay, A. (1995) 'Concepts of equity, fairness and justice in geographical studies', *Transactions of the Institute of British Geographers* 20, 4: 500–8.

Hayek, F. (1960) *The Constitution of Liberty*, London: Routledge & Kegan Paul.

Hays, S. (1989) *Beauty, Health and Permanence: Environmental Politics in the United States 1955–1985,* Cambridge: Cambridge University Press.

Headicar, P. and Bixby, B. (1992) *Concrete and Tyres – Local Development Effects of Major Roads – A Case Study of the M40*, London: CPRE.

Healey, P. (1993) 'The communicative turn in planning theory', in F. Fischer and J. Forester (eds) *The Argumentative Turn in Policy Analysis and Planning*, London: UCL Press.

—— (ed.) (1997) *Collaborative Planning: Shaping Places in Fragmented Societies*, Basingstoke: Macmillan.

—— (1998a) 'Collaborative planning in a stakeholder society', *Town Planning Review* 69, 1: 1–21.

—— (1998b) 'Reflections on integration', paper for the Regional Studies Association Annual Conference: *New Lifestyles, New Regions*, November, Newcastle-upon-Tyne: Department of Town and Country Planning, University of Newcastle.

Healey, P. and Shaw, T. (1993) 'Planners, plans and sustainable development', *Regional Studies* 27: 769–76.

—— (1994) 'Changing meanings of "environment" in the British planning system', *Transactions of the Institute of British Geographers* 19, 4: 425–38.

Heclo, H. (1974) *Modern Social Politics in Britain and Sweden*, New Haven, Conn.: Yale University Press.

Her Majesty's Treasury (1997) *Budget 1997: Tax Measures to Help the Environment*, London: HM Treasury.

—— (2000) *A Fair Deal for Transport and the Environment*, pre-Budget report, November 2000, London: HM Treasury.

Heskin, A. (1980) 'Crisis and response: a historical perspective on advocacy planning', *American Planning Association Journal* 46, 1: 50–63.

Hetherington, P. (1999) 'Newbury bypasses land ruling', *Guardian*, 1 May: 14.

Hetherington, P., Ward, D. and Milne, S. (2000) 'Drivers told "not to go out"', *Guardian*, 14 September.

Hicks, J. (1946) *Value and Capital*, Oxford: Oxford University Press.

Hodge, R. and Hardi, P. (1997) 'The need for guidelines: the rationale underlying the Bellagio principles for assessment', in P. Hardi and T. Zadan (eds) *Assessing Sustainable Development: Principles in Practice*, Winnipeg, Canada: International Institute for Sustainable Development, 7–23.

Hodgson, G. (1997) 'Economics, environmental policy and the transcendence of utilitarianism', in J. Foster (ed.) *Valuing Nature: Economics, Ethics and Environment*, London: Routledge, 48–63.

Holdgate, M. (1996) *From Care to Action: Making A Sustainable World*, London: Earthscan.

Holland, A. (1994) 'Natural capital', in R. Attfield and A. Belsey (eds) *Philosophy and the Natural Environment*, Cambridge: Cambridge University Press, 169–82.

—— (1995) 'The assumptions of cost–benefit analysis: a philosopher's view', in K. Willis and J. Corkindale (eds) *Environmental Valuation: New Perspectives*, Wallingford, Oxfordshire: CAB International, 21–38.

—— (1997) 'Substitutability: or, why strong sustainability is weak and absurdly strong sustainability is not absurd', in J. Foster (ed.) *Valuing Nature: Economics, Ethics and Environment*, London: Routledge, 119–34.

—— (1998) 'Environmental valuation: making room for ethics', *UK CEED Bulletin* 53: 15–16.

Holt-Jensen, A. (1997) 'Strategic development planning in western Norway: Hordaland county and the city of Bergen', in P. Healey, A. Khakee, A. Motte and B. Needham (eds) *Making Strategic Spatial Plans: Innovation in Europe*, London: UCL Press, 135–52.

Hornborg, A. (1994) 'Environmentalism, ethnicity and sacred places: reflections on modernity, discourse and power', *Canadian Review of Sociology and Anthropology* 31, 3: 245–67.

House of Lords Select Committee on the European Communities (1994) *Common Transport Policy – Sustainable Mobility* eighth report, Session 1993–94, HL Paper 50, London: HMSO.

House of Lords Select Committee on Sustainable Development (1995) *Report on Sustainable Development*, HL Paper 72, London: HMSO.

Humphreys, D. (1994) 'Minerals in the modern economy', *Minerals Industry International*, July: 23–7.

IEEP (Institute for European Environmental Policy) (1999) *Manual of Environmental Policy: The EU and Britain*, Release 15, Oxford: Elsevier Science.

Ike, P. (1999) 'Inter-provincial divisions in the Netherlands', in P. Fuchs, M. Smith and M. Arthur (eds) *Mineral Planning in Europe*, Nottingham: Institute of Quarrying, 85–98.

Institution of Highways and Transportation (2000) *Revision of PPG 13: Transport*, submission to the DETR, London: Institution of Highways and Transportation.

International Chamber of Commerce (1992) *From Ideas to Action: Business and Sustainable Development*, Gyldendal, Norway: ICC/Ad Notam.

Irwin, A. and Wynne, B. (1996) *Misunderstanding Science: The Public Reconstruction of Science and Technology*, Cambridge: Cambridge University Press.

IUCN, UNEP and WWF (1991) *Caring for the Earth: A Strategy for Sustainable Living*, Gland, Switzerland: IUCN.

Jachtenfuchs, M. (1996) *International Policy-Making as a Learning Process?* Aldershot: Avebury.

Jachtenfuchs, M. and Huber, M. (1993) 'Institutional learning in the European Community: the response to the greenhouse effect', in J. Liefferink, P. Lowe and A. Mol (eds) *European Integration and Environmental Policy*, London and New York: Belhaven Press.

Jacobs, M. (1991) *The Green Economy*, London: Pluto Press.

—— (1993) *Sense and Sustainability*, London: CPRE.

—— (1995) 'Sustainable development, capital substitution and economic humility: a response to Beckerman', *Environmental Values* 4, 1: 57–68.

—— (1997a) 'Environmental valuation, deliberative democracy and public decision-making institutions', in J. Foster (ed.) *Valuing Nature: Economics, Ethics and Environment*, London: Routledge, 211–31.

—— (1997b) *Making Sense of Environmental Capacity*, London: CPRE.

—— (1997c) 'Introduction: the new politics of the environment', in M. Jacobs (ed.) *Greening the Millennium: The New Politics of the Environment* Oxford: Basil Blackwell, 1–17.

—— (1999) 'Sustainable development as a contested concept', in A. Dobson (ed.) *Fairness and Futurity: Essays on Sustainability and Social Justice*, Oxford: Oxford University Press, 21–45.

Jänicke, M. and Weidner, H. (eds) (1997) *National Environmental Policies: A Comparative Study of Capacity-Building*, Berlin: Springer.

Janssens, J. and van Tatenhove, J. (2000) 'Green planning: from sectoral to integrative planning arrangements?' in J. van Tatenhove, B. Arts and P. Leroy (eds) *Political Modernisation and the Environment: The Renewal of Environmental Policy Arrangements*, Dordrecht: Kluwer, 145–74.

Jemmett, A. (1995) *The Dee Estuary Strategy*, consultation draft, Birkenhead: Wirral Metropolitan Borough Council.

Jessop, B. (1997a) 'Governance of complexity and the complexity of governance: preliminary remarks on some problems and limits of economic guidance', in A. Amin and J. Hausner (eds) *Beyond Market and Hierarchy: Interactive Governance and Social Complexity*, Cheltenham: Edward Elgar, 111–47.

—— (1997b) 'The future of the nation state: erosion or reorganization? General reflections on the West European case', mimeo, Lancaster: Department of Social Science, University of Lancaster.

Jewell, T. (1995) 'Planning regulation and environmental consciousness: some lessons from minerals?' *Journal of Planning and Environment Law*: 482–98.

Jewell, T. and Steele, J. (1996) 'UK regulatory reform and the pursuit of "sustainable development": the Environment Act 1995', *Journal of Environmental Law*: 283–300.

Johnson, L. (1991) *A Morally Deep World: Essays on Moral Significance and Environmental Ethics*, Cambridge: Cambridge University Press.

Kay, P. (1999) 'City to fight curbs on cars', *Sheffield Telegraph* No. 529, 19 November, 1 & 5.

Keat, R. (1997) 'Values and preferences in neo-classical environmental economics', in J. Foster (ed.) *Valuing Nature: Economics, Ethics and Environment*, London: Routledge, 32–47.

Kellett, J. (1995) 'The elements of a sustainable aggregates policy', *Journal of Environmental Planning and Management* 38: 569–79.

Kelman, S. (1990) 'Cost–benefit analysis: an ethical critique', *Regulation* 5, 1: 33–40.

Kemp, R. (1990) '"Why not in my backyard?" A radical interpretation of public opposition to the deep disposal of radioactive waste in the United Kingdom', *Environment and Planning A* 22: 1239–58.

Kennedy, W. (1988) 'Environmental impact assessment in North America and Western Europe: what has worked where, why and how?' *International Environmental Reporter* 11, 4: 257–62.

Kent County Council (1993) *Kent Structure Plan – Third Review: Deposit Plan and Explanatory Memorandum*, Maidstone: Kent County Council.

King, D. and Stoker, G. (eds) (1996) *Rethinking Local Democracy*, Basingstoke: Macmillan.

Kingdon, J. (1995), *Agendas, Alternatives and Public Policies* (second edition), New York and Harlow: Longman.

Kitchen, T. (1997) *People, Politics, Policies and Plans*, London: Paul Chapman.

Kleven, T. (1996) 'Environment and planning – norms and realities', *Scandinavian Housing and Planning Research* 13: 129–46.

Labour Party (1996) *Consensus for Change: Labour's Transport Strategy for the 21st Century*, London: Labour Party.

Lai, L. (1999) 'Hayek and town planning: a note on Hayek's views towards town planning', *Environment and Planning A* 31, 9: 1567–82.

Land Use Consultants (1999) *The Use of the Land Use Planning System to Achieve Non-Land Use Policy Objectives*, report prepared for the Royal Commission on Environmental Pollution, London: Land Use Consultants.

Laslett, P. and Fishkin, J. (eds) (1992) *Justice between Age Groups and Generations*, New Haven, Conn., and London: Yale University Press.

Laslett, R. (1995) 'The assumptions of cost–benefit analysis', in K. Willis and J. Corkindale (eds) *Environmental Valuation: A New Perspective*, Wallingford, Oxfordshire: CAB International, 5–20.

LaViolette, P. and McIntosh, A. (1997) 'Fairy hills: merging heritage and conservation', *ECOS* 18, 3/4: 2–8.

Lawson, T. (1997) '*Beyond 2000* – will it deliver?' *ECOS* 18, 3/4: 57–60.

Leicestershire County Council, Leicester City Council and Rutland County Council (1998) *Leicestershire, Leicester and Rutland Structure Plan Consultation Draft*, Leicester: Leicestershire County Council.

Le-Las, W. (1999) 'Sustaining biodiversity: the contribution of the planning system in controlling development', in N. Herbert-Young (ed.) *Law, Policy and Development in the Rural Environment*, Cardiff: University of Wales Press.

Leopold, A. (1949) *A Sand County Almanac and Sketches Here and There*, Oxford: Oxford University Press.

Levett, R. (1998) 'Focus on: housing allocation', *Environmental Capital Update*, autumn, London: CAG Consultants, 7.

—— (1999a) 'Lessons from the pilots', *EGextra* November/December: 4.

—— (1999b) *Environmental Planning, People's Values and Sustainable Development*, background paper for the Royal Commission on Environmental Pollution, London: CAG Consultants.

Liberal Democrats (1995) *Transporting People: Tackling Pollution*, London: Liberal Democrats.

Lichfield, N. (1994) 'Community impact evaluation', *Planning Theory* 12: 55–79.

Link Quarry Group (1996) *The Case Against the Harris Superquarry*, Edinburgh: Friends of the Earth Scotland.

Litfin, K. (1994) *Ozone Discourses*, New York: Columbia University Press.

Littlewood, S. and While, A. (1997) 'A new agenda for governance? Agenda 21 and the prospects for holistic local decision-making', *Local Government Studies* 23, 4: 111–23.

Lock, D. (1999) 'The new Catch 22 on "need"', *Town and Country Planning* 68, 4: 111.

Locke, J. (1988) 'Second treatise on civil government', in P. Laslett (ed.) *Two Treatises on Government*, Cambridge: Cambridge University Press.

London Borough of Hounslow (1991) *Environmental Charter*, Hounslow: London Borough of Hounslow.

London Economics (1998) *The Environmental Costs and Benefits of the Supply of Aggregates*, London: DETR.

—— (1999) *The Environmental Costs and Benefits of the Supply of Aggregates: Phase 2*, London: DETR.

Low, N. (1994) 'Planning and Justice', in H. Thomas (ed.) *Values and Planning*, Aldershot: Ashgate, 116–39.

Low, N. and Gleeson, B. (1998) *Justice, Society and Nature: An Exploration of Political Ecology*, London: Routledge.

—— (1999) 'Geography, justice and the limits of rights', in J. Proctor and D. Smith (eds) *Geography and Ethics: Journeys in a Moral Terrain*, London: Routledge, 30–43.

Lowe, P. and Goyder, J. (1983) *Environmental Groups in Politics*, London: Allen & Unwin.

Lowe, P., Murdoch, J., Marsden, T., Munton, R. and Flynn, A. (1993) 'Regulating the new rural spaces: the uneven development of land', *Journal of Rural Studies* 9: 205–22.

Lynch, K. (1960) *The Image of the City*, Cambridge, Mass.: MIT Press.

Macalister, T. and Elliott, L. (2000) 'Accused of collusion, firms risk wrath of government', *Guardian*, 14 September: 4.

McAuslan, P. (1979) 'The ideologies of planning law', *Urban Law and Policy* 2: 1–23.

MacEwen, M. and MacEwen, A. (1982) *National Parks: Conservation or Cosmetics?* London: Allen & Unwin.

Mackenzie, A. (1998) '"The Cheviot, the Stag … and the White, White Rock?": Community, identity, and environmental threat on the Isle of Harris', *Environment and Planning D: Society and Space* 16: 509–32.

McKenzie Hedger, M. (1995) 'Wind power: challenges to planning policy in the UK', *Land Use Policy* 12: 17–28.

McKinsey Global Institute (1998) *Driving Productivity and Growth in the UK Economy*, London: McKinsey Global Institute.

McLaren, D. and Bosworth, T. (1994) *Planning for the Planet: Sustainable Development Policies for Local and Strategic Plans*, London: Friends of the Earth.

Macnaghten, P. and Urry, J. (1998) *Contested Natures*, London: Sage.

Macnaghten. P., Grove-White. R., Jacobs, M. and Wynne, B. (1995) *Public Perceptions and Sustainability in Lancashire: Indicators, Institutions, Participation*, a report commissioned by Lancashire County Council, Lancaster: Centre for the Study of Environmental Change, Lancaster University.

McQuail, P. (1994) *Origins of the Department of the Environment*, London: DoE.

Maguire, K. (2000) 'Ten questions: the oil companies confronted', *Guardian* 15 September.

Majone, G. (1976) 'Choice among policy instruments for policy control', *Policy Analysis* 2, 4: 589–613.

—— (1989), *Evidence, Argument and Persuasion in the Policy Process*, New Haven, Conn., and London: Yale University Press.

—— (ed.) (1996) *Regulating Europe*, New York: Routledge.

Marquand, D. (1988) *The Unprincipled Society: New Demands and Old Politics*, London: Fontana Press.

Marshall, T. (1992) *Environmental Sustainability: London's Unitary Development Plans and Strategic Planning*, Occasional Paper 4/1992, London: Faculty of the Built Environment, South Bank University.

Martin, R. and Abercrombie, I. (1995) *Closing Submissions for Redland Aggregates Limited: Lingerbay Inquiry*, unpublished inquiry document, Groby, Leicestershire: Redland Aggregates Ltd.

Martinez-Alier, J., Munda, G. and O'Neill, J. (1998) 'Weak comparability of values as a foundation for ecology and economics', *Ecological Economics* 26, 3: 277–86.

Marvin, S. and Guy, S. (1997) 'Creating myths rather than sustainability: the transition fallacies of the new localism', *Local Environment* 2, 3: 311–20.

Mawhinney, B. (1995) *Transport: The Way Ahead*, London: DoT.

May, P. (1992) 'Policy learning and failure', *Journal of Public Policy* 12, 4: 331–54.

Mead, A. (1999) *West Sussex Minerals Local Plan: Report on Objections*, Bristol: Planning Inspectorate.

Meadowcroft, J. (1997) 'Planning, democracy and the challenge of sustainable development', *International Political Science Review* 18, 2: 167–89.

Meine, C. (1992) 'Conservation biology and sustainable societies: a historical perspective', in M. Oelschlaeger (ed.) *After Earth Day: Continuing the Conservation Effort*, Denton, Texas: University of North Texas Press, 37–65.

Memon, P. and Gleeson, B. (1995) 'Towards a new planning paradigm? Reflections on New Zealand's Resource Management Act', *Environment and Planning B: Planning and Design* 22: 109–24.

Middleton, N. (1999) *The Global Casino: An Introduction to Environmental Issues* (second edition), London: Edward Arnold.

Miles, D. (2000) 'Housing in the south east of England: some issues raised by the government's plans', *World Economics* 1, 2: 1–11.

Miller, C. (1993) 'Coal, smoke and national sovereignty: a case study of the role of planning in controlling pollution', *Journal of Environmental Planning and Management* 36, 2: 149–66.

—— (1999a) *Planning and Pollution Revisited*, report prepared for the Royal Commission on Environmental Pollution, Salford: Academic Enterprise, University of Salford.

—— (1999b) 'Environmental law: the weak versus the strong', *Environmental Law Review*: 23–36.

Miller, C. and Wood, C. (1983) *Planning and Pollution*, Oxford: Oxford University Press.

Mineral Planning (2000) 'New aggregates association', 82, 4.

Mol, A. (1996) 'Ecological modernisation, institutional reflexivity: environmental reform in the late modern age', *Environmental Politics* 5, 2: 302–23.

Moroni, S. (1994) 'Planning, assessment and utilitarianism. Notes on Nathaniel Lichfield's contribution to the evaluation field', *Planning Theory* 12: 81–107.

Municipality of Schiedam (1991) *Physical Planning and Ecology: The Spaland Area*, entry for International Council for Local Environmental Initiatives competition, Schiedam, the Netherlands: Municipality of Schiedam.

Munton, R. (1999) *Environmental Governance and the Politics of Difference*, unpublished address to the Royal Geographical Society/Institute of British Geographers Conference, Leicester, January, London: Department of Geography, University College London.

Murdoch, J. (2000) 'Space against time: competing rationalities in planning for housing', *Transactions of the Institute of British Geographers* 25: 503–19.

Murdoch, J. and Marsden, T. (1995) 'The spatialization of politics: local and national actor-spaces in environmental conflict', *Transactions of the Institute of British Geographers* 20: 368–80.

Murdoch, J. and Tewdwr-Jones, M. (1999) 'Planning and the English regions: conflict and convergence amongst the institutions of regional governance', mimeo, Cardiff: Department of City and Regional Planning, Cardiff University.

Myerson, G. and Rydin, Y. (1994) '"Environment" and planning: a tale of the mundane and the sublime', *Environment and Planning D: Society and Space* 12: 437–52.

—— (1996) *The Language of Environment: A New Rhetoric*, London: UCL Press.

National Rivers Authority (1994) *Guidance Notes for Local Planning Authorities in the Methods of Protecting the Water Environment Through Development Plans*, Bristol: NRA.

National Trust (1999) *A Call for the Wild*, Cirencester: National Trust.

Neal, N. (1996) 'Reserve "will help to preserve threatened wildlife"', *Western Mail: Country Supplement*, 17 September.

Netherlands Ministry of Housing, Spatial Planning and the Environment (1989) *National Environmental Policy Plan: To Choose or to Lose*, The Hague: Ministry of Housing, Spatial Planning and the Environment.

—— (1990) *National Environmental Policy Plan Plus 1990–1994*, The Hague: Ministry of Housing, Spatial Planning and the Environment.

—— (1991a) *Environmental Policy in the Netherlands*, The Hague: Ministry of Housing, Spatial Planning and the Environment.

—— (1991b) *The Right Business in the Right Place*, The Hague: Ministry of Housing, Spatial Planning and the Environment.

—— (1994) *National Environmental Policy Plan 2: The Environment: Today's Touchstone*, summary, The Hague: Ministry of Housing, Spatial Planning and the Environment.

—— (1998) *National Environmental Policy Plan 3* (summary available in English), The Hague: Ministry of Housing, Spatial Planning and the Environment.

Netherlands Ministry of Housing, Spatial Planning and the Environment, Ministry of Transport, Public Works and Ministry of Economic Affairs (1994) *Location Policy in Progress: The Story So Far*, Ref. 31220, The Hague: Ministry of Housing, Spatial Planning and the Environment.

Netherlands Ministry of Transport, Public Works and Water Management (1999) *Perspectievennota Verkeer en Vervoer*, The Hague: Ministry of Transport, Public Works and Water Management.

Newman, P. and Kenworthy, J. (1989) 'Gasoline consumption and cities: a comparison of US cities with a global survey', *Journal of the American Planning Association* 55, 1: 24–37.

Newport County Borough Council (1998) *Newport Unitary Development Plan 1996–2011*, consultation draft, Newport: Planning Department, Newport County Borough Council.

Norgaard, R. (1988) 'Sustainable development: a co-evolutionary view', *Futures*: 606–20.

—— (1994) *Development Betrayed: The End of Progress and a Coevolutionary Revisioning of the Future*, London: Routledge.

Northcott, M. (2000) 'Sabbaths, shamans and superquarrying on a Scottish island: religio-cultural resistance to development in a contested landscape', in F. Gale and M. M'Gonigle (eds) *Nature, Production and Power: Towards an Ecological Political Economy*, Northampton: Edward Elgar, 17–34.

Norton, B. (1982) 'Environmental ethics and the rights of future generations', *Environmental Ethics* 4/winter: 319–37.

Norwegian Geological Survey (1997) *Norway's Coastal Aggregates: Current Production and Potential*, Trondheim: NGU.

Nowell, T. (1991) *SSSIs: A Health Check*, London: Wildlife Link.

O'Neill, J. (1993) *Ecology, Policy and Politics: Human Well-Being and the Natural World*, London: Routledge.

—— (1995) 'Public choice, institutional economics, environmental goods', *Environmental Politics* 4, 2: 197–218.

—— (1998) *The Market: Ethics, Knowledge and Politics*, London: Routledge.

O'Neill, O. (1996) *Towards Justice and Virtue: A Constructive Account of Practical Reasoning*, Cambridge: Cambridge University Press.

—— (1997) 'Environmental values, anthropocentrism and speciesism', *Environmental Values* 6, 2: 127–42.

—— (1998) 'Necessary anthropocentrism and contingent speciesism', comment on A. Wood, 'Kant on duties regarding non-rational nature' (same volume), *Proceedings of the Aristotelian Society*, Supplementary Volume LXXII: 211–28.

O'Sullivan, J., Pritchard, D. and Gammell, A. (1993) 'Saving Europe's wildlife? The EC Habitats Directive', *RSPB Conservation Review* 7: 61–6.

Ove Arup and Partners (1995) *Implementation of PPG 13, 1994–96*, London: DoE.

—— (1999) *The Effectiveness of PPG 13: A Pilot Study*, London: DETR.

Owens, S. (1985) 'Energy, participation and planning: the case of electricity generation in the United Kingdom', in F. Calzonetti and B. Soloman (eds) *Geographical Dimensions of Energy*, Dordrecht: Reidel, 225–53.

—— (1986) *Energy, Planning and Urban Form*, London: Pion.

—— (1991) 'Energy efficiency and sustainable land use patterns', *Town and Country Planning*, February: 44–5.

—— (1992) 'Land use planning for energy efficiency', *Applied Energy* 43: 81–114.

—— (1993) 'Planning and nature conservation – the role of sustainability', ECOS 14, 3&4: 15–22.

—— (1994) 'Land, limits and sustainability: a conceptual framework and some dilemmas for the planning system', *Transactions of the Institute of British Geographers* NS 19: 439–56.

—— (1995) 'Predict and provide or predict and prevent? Pricing and planning in transport policy', *Transport Policy* 2, 1: 43–9.

—— (1997) 'Negotiable environments: needs, demands and values in the age of sustainability', *Environment and Planning A* 29: 571–80.

—— (2000) ' "Engaging the public": information and deliberation in environmental policy', *Environment and Planning A* 32: 1141–8.

Owens, S. and Cope, D. (1992) *Land Use Planning Policy and Climate Change*, London: HMSO.

Owens, S. and Cowell, R. (1994) 'Lost land and limits to growth: conceptual problems for sustainable land use change', *Land Use Policy* 11, 3: 168–80.

—— (1996) *Rocks and Hard Places: Mineral Resource Planning and Sustainability*, London: CPRE.

Owens, S. and Rayner, T. (1999) 'When knowledge matters: the role and influence of the Royal Commission on Environmental Pollution', *Journal of Environmental Policy and Planning* 1, 1: 7–24.

Paehlke, R. and Togerson, D. (eds) (1990) *Managing Leviathan: Environmental Politics and the Administrative State*, London: Belhaven Press.

Pain, G. (1999) *Report of Public Local Inquiry into Extraction of Anorthosite at Lingerbay, Isle of Harris*, Edinburgh: Inquiry Reporters Unit.

Parfit, D. (1984) *Reasons and Persons*, Oxford: Oxford University Press.

Parker, J. (1995) 'Enabling morally reflective communities: towards a resolution of the democratic dilemma of environmental values in policy', in Y. Guerrier, N. Alexander, J. Chase and M. O'Brien (eds) *Values and the Environment: A Social Science Perspective*, Chichester: John Wiley & Sons, 33–49.

Parson, E. and Clark, W. (1997) 'Learning to manage global environmental change: a review of relevant theory', paper presented at symposium, *Innovation of Environmental Policy*, University of Bologna, Italy, 21–25 July, Cambridge, Mass.: Center for Science and International Affairs, Kennedy School of Government, Harvard University.

Partidario, M. and Clark, R. (eds) (2000) *Perspectives on Strategic Environmental Assessment*, London: Lewis Publishers.

Partridge, E. (1984) 'Nature as a moral resource', *Environmental Ethics* 4: 175–90.

Pearce, D. (1988) 'Economics, equity and sustainable development', *Futures* 20, 6: 598–605.

—— (1993) *Blueprint 3: Measuring Sustainable Development*, London: Earthscan.

Pearce, D. and Markandya, A. (1988) *Environmental Considerations and the Choice of Discount Rate in Developing Countries*, World Bank Environment Department Working Paper No. 3, Washington: World Bank.

Pearce, D. and Turner, R. (1990) *Economics of Natural Resources and the Environment*, Hemel Hempstead: Harvester Wheatsheaf.

Pearce, D., Markandya, A. and Barbier, E. (1989) *Blueprint for a Green Economy*, London: Earthscan.

Pennington, M. (1999) 'Free market environmentalism and the limits of land use planning', *Journal of Environmental Policy and Planning* 1, 1: 43–59.

Perks, W. and Tyler M. (1991) 'The challenge of sustainable development: planning for change or changing planning?' *Plan Canada* (Journal of the Canadian Institute of Planners) 31, 3: 6–14.

Peterborough Environment City Trust (1995) *The Peterborough Natural Environment Audit*, Peterborough: Peterborough Environment City Trust.

Petts, J. (1999) 'Public participation and environmental impact assessment' in J. Petts (ed.) *Handbook of Environmental Impact Assessment*, Vol. 1: Oxford: Blackwell Science.

Pezzey, J. (1989) 'Greens and growth – a reply', *UK CEED Bulletin* 22: 22–3.

—— (1992) *Sustainable Development Concepts: An Economic Analysis*, World Bank Environment Paper No. 2, Washington: World Bank.

Phillips, A. (1996) 'The challenge of restoring Europe's nature and landscapes', *International Planning Studies* 1: 73–94.

Plant, J., Turner, K. and Highley, D. (1998) in Proceedings of Minerals '98 Conference, *Minerals, Land and the Natural Environment: The Foundations of Wealth*, 24 June, London: Institution of Mining and Metallurgy.

Plowden, B. (1994) 'Sustainability criteria for minerals planning', *ECOS* 13, 4: 23–6.

Pollock, D. (1999) 'The review of MPG 6 and related issues: a QPA view', *Mineral Planning* 81: 14.

Pollock, S. and Henry, J.J. (1996) 'Mineral extraction and the United Kingdom policies for sustainable development', *Mineral Industry International*, January: 13–16.

Pound, D. (1997) 'Consensus building and nature conservation', *Nature's Place* 15: 7–9.

Pritchard, D. (1994) 'Towards sustainability in the planning system: the role of EIA', *ECOS* 14 (3–4), 10–14.

Pullen, S. (1994) 'Cardiff Bay and the fate of 4000 doomed dunlins', *WWF Marine Update* 16: 1–4.

Pulteney, C. (1999) 'SOS saves SSSI', *Nature's Place* 19: 11–12.

Punter, J. and Carmona, M. (1997) 'Cosmetics or critical constraints? The role of landscape in design policies in English development plans', *Journal of Environmental Planning and Management* 40, 2: 173–97.

Purdue, M. (1999) 'The relationship between development control and specialist pollution controls: which is the tail and which is the dog?' *Journal of Planning and Environment Law*: 585–94.

Quinn, M. (1996) 'Central government planning policy', in M. Tewdwr-Jones (ed.) *British Planning Policy in Transition*, London: UCL Press, 16–28.

RAC Foundation for Motoring and the Environment (1992) *Cars and the Environment: A View to the Year 2020*, London: RAC Foundation.

Radaelli, C. (1995) 'The role of knowledge in the policy process', *Journal of European Public Policy* 2, 2: 159–83.

Rawls, J. (1972) *A Theory of Justice*, Oxford: Oxford University Press.

—— (1993) *Political Liberalism*, New York: Columbia University Press.

Rayner, T. (2001) 'Policy, discourse and institutional reform in the UK transport sector', paper presented at European Consortium for Political Research 29th Joint Session of Workshops, Grenoble, 6–11 April, Cambridge: St John's College.

Raynsford, N. (2000) 'Government seeks view on sustainable aggregates supply', press release, 23 October, London: DETR.

RCEP (Royal Commission on Environmental Pollution) (1994) *Transport and the Environment*, eighteenth report, Cmnd 2674, London: HMSO.

—— (1997) *Transport and the Environment: Developments since 1994*, twentieth report, Cmnd 3752, London: HMSO.

—— (1998) *Setting Environmental Standards*, twenty-first report, Cmnd 4053, London: Stationery Office.

—— (2000) *Energy: The Changing Climate*, twenty-second report, Cmnd 4749, London: Stationery Office.

Reade, E. (1982) 'Section 52 and corporatism in planning', *Journal of Planning and Environment Law*: 8–16.

Redclift, M. (1987) *Sustainable Development: Exploring the Contradictions*, London: Routledge.

Reed, M. and Slaymaker, O. (1993) 'Ethics and sustainability: a preliminary perspective', *Environment and Planning A* 25: 723–39.

Rees, J. (1988) 'Pollution control objectives and the regulatory framework', in R. Turner (ed.) *Sustainable Environmental Management: Principles and Practice*, London: Belhaven Press, 170–89.

—— (1990) *Natural Resources: Allocation, Economics and Policy* (second edition), London: Routledge.

Rees, J. and Williams, S. (1993) *Water for Life: Strategies for Sustainable Water Resource Management*, report prepared for the Council for the Protection of Rural England, London: CPRE.

Rees, W. (1999) 'Scale, complexity and the conundrum of sustainability', in M. Kenny and J. Meadowcroft (eds) *Planning Sustainability*, London: Routledge, 101–27.

Regan, T. (1981) 'The nature and possibility of an environmental ethic', *Environmental Ethics* 3, Spring: 19–34.

—— (1983) *The Case for Animal Rights*, Berkeley: University of California Press.

Rein, M. and Schön, D. (1991) 'Frame-reflective policy discourse', in P. Wagner, C. Weiss, B. Wittrock and H. Wollman (eds) *Social Sciences and Modern States*, Cambridge: Cambridge University Press, 262–89.

Richardson, N. (1989) *Land Use Planning and Sustainable Development in Canada*, Ottowa: Canadian Environmental Advisory Council.

Richardson, T. (1996) 'Foucaldian discourse: power and truth in urban and regional policy making', *European Planning Studies* 4, 3: 279–92.

Richardson, T. and Haywood, R. (1996) 'Deconstructing transport planning: lessons from policy breakdown in the English Pennines', *Transport Policy* 3, 1/2: 43–53.

Ritsema, G. and Asselman, G. (1994) *Stop de asfaltering: vier jaar verkeersbeleid onder de loep*, brochure, Amsterdam: Vereniging Milieudefensie.

Roberts, R. and Emel, J. (1992) 'Uneven development and the tragedy of the commons: competing images for nature–society analysis', *Economic Geography* 68: 249–71.

RSPB (Royal Society for the Protection of Birds) (1993) *Response to the UK Strategy for Sustainable Development Discussion Paper*, Sandy, Bedfordshire: RSPB.

—— (1996) 'European Court backs wildlife site protection', *Conservation Planner* 7: 1.

—— (1997) *The Nature of Regions: Regional Planning and Nature Conservation in the UK*, Sandy, Bedfordshire: RSPB.

—— (1999) *Implementing Biodiversity Action Plans*, proceedings of a one-day conference hosted by Government Office for the East of England (ed. Darren Kinleysides), RSPB Central England Office, 46 The Green, South Bar, Banbury, OX16 9AB, UK.

—— (2000) 'The Countryside Bill: what it means for planners', *Conservation Planner* 14: 1.

RTPI (Royal Town Planning Institute) (1996) *Energy Planning: A Guide for Practitioners*, London: RTPI.

—— (1999) *Planning for Biodiversity*, Peterborough: English Nature.

Rural Development Commission (1995) *Planning for People and Prosperity*, London: Rural Development Commission.

Rydin, Y. (1995) 'Sustainable development and the role of land use planning', *Area* 27, 4: 369–77.

—— (1998a) *The British Planning System: An Introduction* (second edition), Basingstoke: Macmillan.

—— (1998b) 'Land use planning and environmental capacity: reassessing the use of regulatory policy tools to achieve sustainable development', *Journal of Environmental Planning and Management* 41, 6: 749–65.

Sabatier, P. (1987) 'Knowledge, policy-oriented learning and policy change: an advocacy coalition framework', *Knowledge: Creation, Diffusion, Utilisation* 8, 4: 649–92.

—— (1998) 'The advocacy coalition framework: revisions and relevance for Europe', *Journal of European Public Policy* 5, 1: 98–130.

Sabatier, P. and Jenkins Smith, H. (eds) (1993) *Policy Change and Learning: An Advocacy Coalition Approach*, Boulder, Col.: Westview Press.

Sabatier, P. and Mazmanian, D. (1979) 'The conditions of effective implementation: a guide to accomplishing policy objectives', *Policy Analysis* 5, 4: 481–504.

SACTRA (1994) *Trunk Roads and the Generation of Traffic*, London: HMSO.

—— (1999) *Transport and the Economy*, London: Stationery Office.

Sadler, B. (1996) *Environmental Assessment in a Changing World: Evaluating Practice to Improve Performance*, International Study of the Effectiveness of Environmental Assessment, final report, Ottowa: Canadian Environmental Assessment Agency and International Association for Impact Assessment.

Sagoff, M. (1981) 'Do we need a land use ethic?' *Environmental Ethics* 3, winter: 293–308.

—— (1988) *The Economy of the Earth*, Cambridge, Cambridge University Press.

Sandbach, F. (1980) *Environment, Ideology and Policy*, Oxford: Basil Blackwell.

Saward, M. (1993) 'Green democracy', in A. Dobson and P. Lucardie (eds) *The Politics of Nature: Explorations in Green Political Theory*, London: Routledge, 63–80.

Sayer, R. (1979) 'Understanding urban models versus understanding cities', *Environment and Planning A* 11: 853–62.

Schattschneider, E. (1960) *The Semisovereign People*, New York: Holt, Rinehart & Wilson.

Scottish Biodiversity Group (1997) *Biodiversity in Scotland: The Way Forward*, Edinburgh: Scottish Biodiversity Group.

Scottish Office (1997) *Local Biodiversity Action Plans: A Manual*, Edinburgh: Scottish Office.

Scottish Office Development Department (1999) National Planning Policy Guideline 15: *Rural Development*, Edinburgh: Scottish Office.

ScottishPower (1999) *Environment Report 1998–99*, Glasgow: ScottishPower.

Searle, G. (1975) 'Copper in Snowdonia', in P. Smith (ed.) *The Politics of Physical Resources*, London: Open University Press, 66–112.

—— (1980) 'Copper in Snowdonia National Park', in P. Smith (ed.) *The Politics of Physical Resources*, Harmondsworth: Penguin, 66–112.

Selman, P. (1995) Local sustainability: can the planning system help us to get from here to there?' *Town Planning Review* 66, 3: 287–301.

—— (1996) *Local Sustainability: Managing and Planning Ecologically Sound Places*, London: Paul Chapman.

—— (2001) 'Social capital, sustainability and environmental planning', *Planning Theory and Practice* 2, 1: 13–30.

Selman, P. and Wragg, A. (1999) 'Local sustainability planning: from interest-driven networks to vision-driven super-networks?' *Planning Practice and Research* 14, 3: 329–40.

Sen, A. (1987) *On Ethics and Economics*, Oxford: Basil Blackwell.

—— (1999) 'The reach of consequential valuation', text relating to Heffer Lecture, 'Rights, Duties and Consequences', University of Cambridge Faculty of Philosophy, November 1998, Cambridge: Trinity College.

SERPLAN (London and South East Regional Planning Conference) (1999) *A Sustainable Development Strategy for the South East*, London: SERPLAN.

Shepherd, P. and Gillespie, J. (1996) *Developing Definitions of Natural Capital for Use within the Uplands of England*, Research Report No. 197, Peterborough: English Nature.

Shepherd, P. and Harley, D. (1999) *Preparation and Presentation of Habitat Replacement Costs Estimates*, Peterborough: English Nature.

Shrader-Frechette, K. (1985) *Science Policy, Ethics, and Economic Methodology*, Dordrecht: Kluwer.

Simpson, B. (1987) *Planning and Public Transport in Great Britain, France and West Germany*, London: Longman.

Singer, P. (1976) *Animal Liberation*, London: Jonathan Cape.

—— (1979) *Practical Ethics*, Cambridge: Cambridge University Press.

Singer, P. and Cavalieri, P. (1993) *The Great Apes Project*, London, Fourth Estate.

Skolimovski, H. (1995) 'In defence of sustainable development', *Environmental Values* 4, 1: 69–70.

Slower Speeds Initiative (1998) *Policy Briefing No. 1*, Norwich: Slower Speeds Initiative.

Smith, D. (1994) 'On professional responsibility to distant others', *Area* 26: 359–67.

Smith, J. (1994) 'The politics of environmental conflict: the case of transport in Great Britain', unpublished PhD thesis, Department of Geography, University of Cambridge.

Smith, N. (1990) *Uneven Development: Nature, Capital and the Production of Space*, Oxford: Basil Blackwell.

SNH (Scottish National Heritage) (1997) *Progress and Plans, 1997*, Perth: SNH.

—— (1998) *Jobs and the National Heritage*, Perth: SNH.

SOEnD (Scottish Office Environment Department) (1994) National Planning Policy Guideline 4: *Land for Mineral Working*, Edinburgh: Crown Copyright.

Solesbury, W. (1976) 'The environmental agenda: an illustration of how situations may become political issues and issues may demand responses from government: or how they may not', *Public Administration* 54, Winter: 379–97.

Solow, R. (1986) 'On the intergenerational allocation of natural resources', *Scandinavian Journal of Economics* 88, 1: 141–9.

—— (1992) 'An almost practical step toward sustainability', fortieth anniversary lecture, Washington: Resources for the Future.

South East Regional Planning Guidance Examination in Public Panel (chair: Professor Stephen Crow) (1999) *Regional Planning Guidance for the South East of England: Report of the Panel,* Guildford: Government Office for the South East.

Southgate, M. (1998) 'Sustainable planning in practice?' *Town and Country Planning,* December: 372–5.

Spash, C. (1998) 'Environmental values and environmental valuation', *UK CEED Bulletin* 53: 12–14.

Spash, C. and Clayton, A. (1995) *Strategies for the Maintenance of Natural Capital,* Discussion Paper in Ecological Economics 95/5, Stirling: Department of Economics, University of Stirling.

Still, B. (1996) 'The importance of transport impacts on land use in strategic planning', *Traffic Engineering and Control,* October: 564–71.

Stirling, A. and Mayer, S. (1999) *Rethinking Risk: A Pilot Multi-Criteria Mapping of a Genetically Modified Crop in Agricultural Systems in the UK,* Brighton: Science and Technology Policy Research, University of Sussex.

Summerton, N. (1998) 'Why water may not stem the housing tide', *Town and Country Planning,* January/February: 26–8.

Surveyor (2000) 'A return to road building?' 13 January: 12–14.

Susskind, L. (1981) 'Environmental mediation and the accountability problem', *Vermont Law Review* 6, 1: 1–47.

—— (2000) 'Mediating environmental disputes: is it really possible to find common ground between tree huggers and avid capitalists?' the Denman Lecture, 15 May, Cambridge: Department of Land Economy, University of Cambridge.

Susskind, L., McKearnan, S. and Thomas-Larmer, J. (1999) *The Consensus Building Handbook: A Comprehensive Guide to Reaching Agreement,* London: Sage.

Sverdrup, L. (1998) 'Local Agenda 21 in Norway', in T. O'Riordan and H. Voisey (eds) *The Transition to Sustainability,* London, Earthscan: 263–75.

Tait, M. and Campbell, H. (2000) 'The politics of communication between planning officers and politicians: the exercise of power through discourse', *Environment and Planning A* 32, 3: 489–506.

Taylor, P. (1986) *Respect for Nature: A Theory of Environmental Ethics,* Princeton, NJ: Princeton University Press.

TCPA (1999) *Your Place and Mine: Reinventing Planning,* Report of the TCPA Inquiry into the Future of Planning, London: TCPA.

Tewdwr-Jones, M. (ed.) (1996) *British Planning Policy in Transition,* London: UCL Press.

—— (2000) review of J. Forester (1999) *The Deliberative Practitioner,* Cambridge, Mass.: MIT Press, *International Planning Studies* 5, 3: 419–21.

Thayer, R. (1994) *Gray World, Green Heart: Technology, Nature and the Sustainable Landscape,* New York: John Wiley & Sons.

Therivel, R. (1995) 'Environmental appraisal of development plans: current status', *Planning Practice and Research* 10, 2: 223–34.

Therivel, R. and Partidario, M. (1996) *The Practice of Strategic Environmental Assessment,* London: Earthscan.

Therivel, R. and Thompson, S. (1996) *Strategic Environmental Assessment and Nature Conservation*, report to English Nature, Peterborough: English Nature.

Therivel, R., Wilson, E., Thompson, S., Heaney, D. and Pritchard, D. (1992) *Strategic Environmental Assessment,* London: Earthscan.

Thomas, H. (ed.) (1994) *Values and Planning*, Aldershot: Avebury.

Thomas, H. and Imrie, R. (1999) 'Urban policy, modernisation, and the regeneration of Cardiff Bay', in R. Imrie and H. Thomas, (eds) *British Urban Policy* (second edition), London: Sage, 106–27.

Tomalty, R. and Hendler, S. (1991) 'Green planning: striving towards sustainable development in Ontario's municipalities', *Plan Canada* (Journal of the Canadian Institute of Planners) 31, 3: 27–32.

Truelove, P. (1999) 'Transport planning', in B. Cullingworth (ed.) *British Planning: 50 Years of Urban and Regional Policy*, London: Athlone Press.

Turner, R. (1988) 'Wetland conservation: economics and ethics', in D. Collard, D. Pearce and D. Ulph, (eds) *Economics, Growth and Sustainable Environments*, Basingstoke: Macmillan, 121–59.

Turner, R., Brouwer, R. and Georgion, S. (2000) 'Ecosystem functions and the implications for environmental valuation', Peterborough: English Nature.

Tyme, J. (1978) *Motorways v. Democracy*, London: Macmillan.

UK Government (1985) *Lifting the Burden*, Cmnd 9571, London: HMSO.

—— (1990) *This Common Inheritance*, London: HMSO.

—— (1994a) *Sustainable Development: The UK Strategy*, London: HMSO.

—— (1994b) *Biodiversity: The UK Action Plan*, Cmnd 2428, London: HMSO.

—— (1995a) *Our Future Homes: Opportunity, Choice, Responsibility*, Cmnd 2901, London: HMSO.

—— (1995b) *Response to the House of Lords Select Committee on Sustainable Development*, London: HMSO.

—— (1996) *Household Growth: Where Shall We Live?* London: Stationery Office.

UK Local Issues Advisory Group (1997a) Guidance Note 1: *Guidance for Local Biodiversity Action Plans: An Introduction*, London: DETR UK Biodiversity Secretariat.

—— (1997b) Guidance Note 2: *Guidance for Local Biodiversity Action Plans: Developing Partnerships*, London: DETR UK Biodiversity Secretariat.

—— (1997c) Guidance Note 3: *Guidance for Local Biodiversity Action Plans: How Local Biodiversity Action Plans Relate to Other Plans*, London: DETR UK Biodiversity Secretariat.

—— (1997d) Guidance Note 4: *Guidance for Local Biodiversity Action Plans: Evaluating Priorities and Setting Targets for Habitats and Species*, London: DETR UK Biodiversity Secretariat.

—— (1997e) Guidance Note 5: *Guidance for Local Biodiversity Action Plans: Incentives and Advice for Biodiversity*, London: DETR UK Biodiversity Secretariat.

UK MAB (Man and the Biosphere) Committee (1998) *Local Nature Reserves: A Time for Reflection: A Time for New Action*, Urban Forum of the UK MAB Committee (Secretary: c/o The Wildlife Trusts, The Green, Witham Park, Waterside South, Lincoln LN5 7JR, UK).

UKRT (UK Round Table on Sustainable Development) (1996a) *Defining a Sustainable Transport Sector*, London: UKRT.

—— (1996b) *Freight Transport*, London: UKRT.

—— (1997a) *Freshwater*, London: UKRT.

—— (1997b) *Housing and Urban Capacity*, London: UKRT.

—— (1997c) *Getting Around Town*, London: UKRT.

—— (1997d) *Making Connections*, London: UKRT.

UNCED (1992a) *Agenda 21: The United Nations Programme of Action from Rio*, New York: United Nations Department of Public Information.

—— (1992b) *Convention on Biological Diversity*, New York: UN Publications.

University of Hull Institute of City and Regional Studies (2000) written evidence submitted to Royal Commission on Environmental Pollution Study of Environmental Planning: http://www.rcep.org.uk

Upton, S. (1995) *Purpose and Principle in the Resource Management Act*, the Stace Hammond Grace Lecture, Waikato, New Zealand: University of Waikato.

Urban Task Force (chair: Lord Rogers of Riverside) (1999) *Towards an Urban Renaissance*, executive summary, London: DETR.

van der Bergh, J. and Verbruggen, H. (1999) 'Spatial sustainability, trade and indicators: an evaluation of the ecological footprint', *Ecological Economics* 29, 61–72.

Van der Gun, V. and de Roo, G. (1994) 'An integrated environmental approach to land-use zoning', in H. Voogd (ed.) *Issues in Environmental Planning*, London: Pion, 58–66.

van der Straaten, J. and Ugelow, J. (1994) 'Environmental policy in the Netherlands: change and effectiveness', in M. Wintle and R. Reeve (eds) *Rhetoric and Reality in Environmental Policy: The Case of the Netherlands in Comparison with Britain*, Aldershot: Avebury, 118–44.

Vanderseypen, G. (1998) 'Mineral planning in Europe: the European Commission's view', in B. van der Moolen, A. Richardson and H. Voogd (eds) *Mineral Planning in a European Context*, Groningen, the Netherlands: Geo Press, 41–8.

Vereniging Milieudefensie (undated) *Minder Schipol, Meer Milieu*, Amsterdam: Vereniging Milieudefensie.

Victor, P.A. (1991) 'Indicators of sustainable development: some lessons from capital theory', *Ecological Economics* 4: 191–213.

Wackernagel, M. and Rees, W. (1996) *Our Ecological Footprint: Reducing Human Impact on the Earth*, Gabriola Island, British Columbia: New Society.

Wackernagel, M., Onisto, L., Bello, P., Linares, A., Falfan, I., Garcia, J., Guerrero, A. and Guerro, M. (1999) 'Natural capital accounting with the ecological footprint concept', *Ecological Economics* 29: 375–90.

Walzer, M. (1983) *Spheres of Justice: A Defence of Pluralism and Equality*, Oxford: Basil Blackwell.

Ward, S. (1993) 'Thinking global, acting local? British local authorities and their environmental plans', *Environmental Politics* 2, 3: 453–78.

WCED (World Commission on Environment and Development) (1987) *Our Common Future*, Oxford: Oxford University Press.

Weale, A. (1992) *The New Politics of Pollution*, Manchester: Manchester University Press.

Webster, F., Bly, P. and Paulley, N. (1988) *Urban Land Use and Transport Interaction: Policies and Models*, report of the International Study Group on Land Use/Transport Interaction, Aldershot: Avebury.

Weiss, C. (1977) 'Research for policy's sake: the enlightenment function of social research', *Policy Analysis* 3, 4: 531–45.

—— (1991) 'Policy research: data, ideas, or arguments?' in P. Wagner, C. Weiss, B. Wittrock and H. Wollman (eds), *Social Sciences and Modern States. National Experiences and Theoretical Crossroads*, Cambridge: Cambridge University Press.

Welbank, M. (1993) *Sustainable Development: A Discussion Paper*, London: RTPI.

Welsh Office (1991) 'Cardiff Bay excluded from the Lower Severn Estuary SPA', press release, 1 November.

—— (1997) *Land Authority for Wales (Gwent Levels, Wetlands Reserve, Newport) Compulsory Purchase Order 1997*, Secretary of State's decision letter, Cardiff: Welsh Office Public Inquiry Unit.

West Sussex County Council (1996) *The Environmental Capacity of West Sussex*, Chichester: West Sussex County Council.

—— (1997) *West Sussex Minerals Local Plan, Deposit Draft: Strategic Environmental Appraisal*, Chichester: West Sussex County Council Planning Department.

Western Isles Labour Party (1994) 'Submission to the public inquiry into the proposed development of a coastal quarry at Lingerbay, Isle of Harris', unpublished inquiry document, Stornoway: WILP.

Whatmore, S. and Boucher, S. (1993) 'Bargaining with nature: the discourse and practice of "environmental planning gain"', *Transactions of the Institute of British Geographers* NS 18: 166–82.

Whitbread, M. and Marsay, A. (1992) *Coastal Superquarries to Supply South-east England Aggregates Requirements*, report by Arup Economics and Planning for the DoE, London: HMSO.

White, C. (1993) 'Values come on styles, which mate to change', in M. Hechter, L. Nadel and R. Michod (eds) *The Origin of Values*, Hawthorne, NY: Aldine de Gruyter, 63–91.

Wiggins, D. (1999) 'Nature, respect for nature and the human scale of values', presidential address, 11 October, *Proceedings of the Aristotelian Society* 100, 1 (September/October): 1–32.

Wildavsky, A. (1993) 'On the social construction of distinctions: risk, rape, public goods and altruism', in M. Hechter, L. Nadel and R. Michod (eds) *Origin of Values*, New York: Aldine de Gruyter.

Williams, R. (1976) *Keywords: A Vocabulary of Culture and Society*, London: Fontana and Croom Helm.

Wilson, E. (1998) 'Planning and environmentalism in the 1990s', in P. Allmendinger and H. Thomas (eds) *Urban Planning and the British New Right*, London: Routledge, 21–52.

Wilson, E. and Raemakers, J. (1992) *Index to Local Authority Green Plans*, Research Paper 44, School of Planning and Housing, Edinburgh: Edinburgh College of Art.

Winter, P. (1994) 'Planning and sustainability: an examination of the role of the planning system as an instrument for the delivery of sustainable development', *Journal of Planning and Environment Law*: 883–900.

Wise, S. (2000) *Rattling the Cage: Towards Legal Rights for Animals*, London: Profile Books.

Wissenburg, M. (1998) *Green Liberalism: The Free and the Green Society*, London: UCL Press.

Wolters, R. and van der Moolen, B. (1999) 'European mineral extraction and the Biodiversity Act', in P. Fuchs, M. Smith and M. Arthur (eds) *Mineral Planning in Europe*, Nottingham: Institute of Quarrying, 12–18.

Wood, C. (1989) *Planning Pollution Prevention*, Oxford: Henemann Newnes.

Wright, P. (1985) *On Living in an Old Country*, London: Verso.

WWF (WorldWide Fund for Nature) (1999) *Natura 2000: Opportunities and Obstacles*, Vienna: WWF.

—— (undated) *The Habitats Directive Scandal*, press information, Brussels: WWF European Policy Office.

WWF UK (1997) *A Muzzled Watchdog: Is English Nature Protecting Wildlife?* Godalming: WWF UK.

Wynne, B. (1975) 'The rhetoric of consensus politics: a critical review of technology assessment', *Research Policy* 4: 108–58.

Yeager, P. (1991) *The Limits of Law: The Public Regulation of Private Pollution*, Cambridge: Cambridge University Press.

Yearley, S. (1996) *Sociology, Environmentalism, Globalization*, London: Sage.

Young, S. (1996) *Promoting Participation and Community-Based Partnerships in the Context of Local Agenda 21: A Report for Practitioners*, Manchester: Government Department, University of Manchester.

INDEX

agency, questions of addressed 63–4
Agenda 21 48, 178n
aggregates: advantages of greater
regional self-sufficiency 148;
Berkshire, unsustainability of
regional guidance 137; effects of
greater procedural fairness in
planning 151–2; ethical
considerations of 'imports' 153;
greater degree of sub-national
flexibility 156; importance of sector
131; large volumes still required
156–7; low-cost and abundant
supplies 131–2; most demands met
locally 155; 'need' justification for
extraction 76; planning for 128;
(becoming more responsive to local
circumstances 154; policies in Wales
and Scotland matters for devolved
administrations 153); pressures to
consider new spatial or structural
fixes 130; primary falling demand for
155–6; proportion from traditional
sources to be reduced 134; proposed
tax on virgin extraction 135; trade
and planning in a transnational
context 142–5; (exporting
unsustainability 143, 144–5); use of
secondary/recycled materials 135,
155, 161, 191n
agriculture: annual payments for
positive management 105, 187n; and
land degradation 105, 186n
air-quality standards 37; and national
air-quality strategy 17–18, 37, 78,
171n, 172n
analytical arms race 50–8, 72; assessing
capacities for development 53–6;

costs, benefits and environmental
capital 56–8; environmental
assessment 50–3
anthropocentric considerations, and
conservation 41
anthropocentrism 176n
appraisal: new approach to (end 1990s)
83; shift towards broader forms of 56
appraisal/audit-type approaches 50–8;
and integration 64–5
argument and persuasion, promoting
conceptions of sustainable
development 164–5
aspirations: difficult to translate into
policy measures 23; Leicester and
Rutland Draft Structure Plan typical
173n; and outcomes, explanations for
gap between 161–2
assets, protected for future generations
35
assumptions, deeply embedded,
challenges to 71

Barksore Marshes, Kent, planning
consent for land reclamation revoked
118
Bedfordshire, precautionary policy on
land allocation not allowed 23
Berkshire, 'green travel' plan wins
company green field location 1, 161
Berkshire County Council, attempts to
revise minerals plan using
sustainability concepts 136–7;
(original objective achieved 138)
Berlin, I. 42–3
biodiversity: challenge of 'no net loss' or

229